MANAGING AND IMPROVING
WAREHOUSE
OPERATIONS

MANAGING AND IMPROVING
WAREHOUSE OPERATIONS

Albert Wee Kwan Tan
Ton Duc Thang University, Vietnam

Siti Norida Wahab
Universiti Teknologi MARA, Malaysia

Huay Ling Tay
Singapore University of Social Science, Singapore

See Ching Chin
Singapore Institute of Materials Management, Singapore

We World Scientific

NEW JERSEY · LONDON · SINGAPORE · BEIJING · SHANGHAI · HONG KONG · TAIPEI · CHENNAI · TOKYO

Published by

World Scientific Publishing Co. Pte. Ltd.

5 Toh Tuck Link, Singapore 596224

USA office: 27 Warren Street, Suite 401-402, Hackensack, NJ 07601

UK office: 57 Shelton Street, Covent Garden, London WC2H 9HE

Library of Congress Control Number: 2025015486

British Library Cataloguing-in-Publication Data
A catalogue record for this book is available from the British Library.

MANAGING AND IMPROVING WAREHOUSE OPERATIONS

ISBN 978-981-98-1274-5 (hardcover)
ISBN 978-981-98-1275-2 (ebook for institutions)
ISBN 978-981-98-1276-9 (ebook for individuals)

For any available supplementary material, please visit
https://www.worldscientific.com/worldscibooks/10.1142/14307#t=suppl

Desk Editors: Murali Appadurai/Pui Yee Lum

Typeset by Stallion Press
Email: enquiries@stallionpress.com

Preface

Warehousing is a cornerstone of modern supply chain management, serving not only as a storage function but also as a critical enabler of speed, accuracy, and customer satisfaction. As global competition intensifies and consumer expectations evolve, the need for efficient, technology-driven, and sustainable warehouse operations has never been more pressing.

The genesis of this book lies in the recognition of key challenges and gaps that persist in warehouse and inventory management. Many organizations struggle with suboptimal layouts, inaccurate inventory tracking, delayed replenishment, and underutilized technologies. We aim to bridge these gaps by presenting a practical, structured, and accessible guide to warehouse optimization.

This book is not just a theoretical exposition; it is a real-world playbook designed to empower warehouse owners, managers, and supply chain professionals with the knowledge and tools necessary to enhance inventory processes, drive efficiency, and improve profitability. The content is structured to provide a comprehensive understanding of the critical facets of warehouse management, from foundational concepts to advanced practices.

Key features of this book include the following:

1. **In-depth Coverage of Core Topics:** Each chapter offers detailed explorations of essential warehouse and inventory management areas, enriched with practical examples and actionable insights.
2. **Emphasis on Sustainable Practices:** Recognizing the growing importance of environmental responsibility, we have dedicated a

chapter to green warehousing, including strategies for reducing carbon footprints and implementing ISO 14000 standards.

3. **Integration of Technology:** This book explores the transformative role of Warehouse Management Systems (WMS), automation, and robotics, discussing both the benefits and challenges of digital adoption.

4. **Focus on Safety and Security:** With warehouse environments becoming more complex, ensuring safety and security is paramount. We provide a comprehensive overview of occupational safety measures, inventory protection protocols, and risk management approaches.

5. **Case Studies and Best Practices:** Real-world case studies and lessons from industry leaders illustrate successful inventory and warehouse strategies, helping readers contextualize concepts and apply them effectively.

6. **Skill Development:** We also address the human element of warehouse operations, offering guidance on staff training, skill enhancement, and organizational capability building.

This book begins with an overview of warehouse operations and types of ownership, setting the stage for deeper discussions on layout design, inventory segmentation, replenishment strategies, and lean practices. Special emphasis is placed on green initiatives, automation, and the importance of continuous improvement through performance metrics and cultural change.

In writing this book, our goal has been to deliver a resource that is both intellectually robust and practically applicable. Whether you are a student, a logistics professional, or a warehouse operator seeking to improve operational outcomes, we hope you find this book insightful and impactful in your journey toward warehouse excellence.

About the Authors

Albert Wee Kwan Tan is an Associate Professor at the Asian Institute of Management, specializing in supply chain management, logistics, and business analytics. He is also a Visiting Professor at Ton Duc Thang University, Vietnam. His academic credentials include a bachelor's degree in Information Technology from the University of Southern Queensland and a master's in Business Studies from the National University of Ireland. Additionally, he holds professional certifications, such as the Certified Fellow in Production and Inventory Management (CFPIM) from APICS.

Dr. Tan has held senior academic positions at prestigious institutions, including Shanghai Jiao Tong University, the National University of Singapore, the MIT SCALE Network, and the University of Wollongong in Dubai. He has also directed academic programs in IT Management, MBA, and MSc Logistics across multiple locations, including Malaysia, Dubai, the Philippines, and Singapore. His research focuses on supply chain and logistics, operations research, and business analytics, with significant contributions in areas, such as optimizing reverse logistics networks, dynamic simulation of warehouse operations, and digital transformation. His work has been published in leading academic journals, and he serves on the editorial boards of the *International Journal of Information Systems In the Service Sector* and the *International Journal of Information Systems and Supply Chain Management*. Dr. Tan's publications can be accessed at Google Scholar.

Siti Norida Wahab is a Senior Lecturer in the operations management program at the Faculty of Business and Management, UiTM Puncak Alam. She has around 20 years of experience in both the industrial and educational fields. Her previous leadership positions include roles at the managerial level in multinational logistics companies and renowned private universities. Her research interest includes sustainable adoption in logistics and supply chain management. She managed national and international grants and published her research works in high-rated journals, proceedings, book and book chapters. Besides, she supervised a number of PhD and MSc candidates. For excellence, she has won platinum, diamond, gold, silver, and bronze medals in innovation competitions. She also actively serves as a reviewer for journals and academic conferences. Currently, she is a professional technologist of the Malaysia Board of Technologists.

Huay Ling Tay is an Associate Professor in Logistics and Supply Chain Management at the Singapore University of Social Sciences (SUSS) School of Business. Her academic background includes a Ph.D. in Business and Economics (Operations Management) from the University of Melbourne, an MSc in Industrial Engineering from Georgia Institute of Technology, an MSc in Logistics and Supply Chain Management, and a BEng (Hon) in Chemical Engineering from the National University of Singapore.

Prior to academia, Dr. Tay gained significant industry experience as a Global Lead for Business Remodelling at APL Ltd, Regional Supply Chain Network Optimisation at Dow Chemicals, and Business Development at Vopak Asia Pte Ltd. She also worked as an Industrial Hygiene Engineer at the Ministry of Manpower Singapore. She teaches operations and supply chain management and Lean Six Sigma. She's also a dedicated pro-bono consultant for non-profits such as Food Bank Singapore and the United Nations World Food Programme, where she improves operations and supply chains through process streamlining. She also contributed to Singapore Standard SS644 on medication supply guidelines.

Dr. Tay's research focuses on Lean Six Sigma, Sustainable Business Values, Healthcare Operations, and Humanitarian Logistics and Supply Chains. She has published and presented numerous articles on operations management and process improvements across various sectors, often applying Lean Six Sigma and performance management frameworks.

Her contributions have been recognized with several awards, including the Best Business Management and Operations Management Paper Award at the IEOM Sydney 2024 Conference, the Global Business Management Award at the IEOM Melbourne Conference 2023, and a First Place in the Lean Six Sigma Competition Paper at the same conference. She also received an Honourable Mention for the SUSS Award for Teaching Excellence 2022.

See Ching Chin holds an MBA in International Business from Australia and a bachelor's degree in Industrial Engineering from Canada. In the early stages of his career, he worked in the manufacturing industry, where he developed expertise in industrial engineering, particularly in motion and time study design and work measurement. He later transitioned to the logistics sector, focusing on Just-In-Time (JIT) hub implementation, transportation management, warehousing, inventory management, and third-party logistics (3PL) solutions. Mr. Chin has gained valuable international experience through postings in Malaysia and China, where he successfully managed cross-cultural relationships by sharing best practices in warehouse management with local teams. This diverse background inspired him to write this book, aimed at providing practical guidance for business owners seeking to enhance their warehouse management practices. Currently, he serves as a Senior Logistics Consultant at his own company, specializing in Logistics and Operations Management as well as Productivity Measurement and Improvement.

Acknowledgment

We would like to express our sincere appreciation to Dickson Suan Liang Yeo for his valuable contribution to the chapter on *Automation and Robotics In Warehousing*. His expertise and practical insights into warehouse automation — from systems integration to operational transformation — have added depth and relevance to this book. Dickson's commitment to advancing warehousing practices through automation, and his ability to bridge theory with real-world application, greatly enhanced the quality and impact of this chapter. We are deeply grateful for the time, knowledge, and collaboration shared throughout this process.

Contents

Chapter 1

Introduction

Learning Outcome

By the end of this chapter, you should be able to do the following:

1. Recognize the warehouse operations.
2. Identify the type of warehouse ownership.
3. Determine the optimal number of warehouses.
4. Explain the centralized vs. decentralized warehousing.
5. Describe the decision-making hierarchy.
6. Clarify the warehousing policies.
7. Decide the importance of effective inventory management.

1. Introduction

In the ever-evolving landscape of logistics and supply chain management, the efficient operation of warehouses and the precise management of inventory have become pivotal factors in ensuring business success. This book, "Optimizing Warehouse Inventory Management," is meticulously crafted to address the widespread inventory challenges faced by user-owned warehouses. Unlike third-party logistics (3PL) companies, which are equipped with sophisticated management systems and expertise, user-owned warehouses often lack the specialized knowledge required for effective inventory control and warehouse operations.

1

The genesis of this book lies in the recognition of these critical gaps. Our aim is to bridge them by presenting fundamental concepts and systematic approaches to enhance warehouse inventory management. This book is not just a theoretical exposition but also a practical guide, offering real-world solutions to common issues. It seeks to empower warehouse owners and managers with the knowledge and tools necessary to optimize their inventory processes, thereby driving efficiency and profitability.

This book is structured to provide a comprehensive understanding of various facets of warehouse and inventory management. Starting with an introduction to warehouse operations and the importance of effective inventory management, it progresses through critical topics, such as warehouse layout and design, inventory classification, replenishment strategies, and green warehousing.

1.1 *Key features of this book*

1. **In-depth coverage of core topics:** Each chapter delves deeply into essential aspects of warehouse and inventory management, providing detailed explanations and practical examples.
2. **Emphasis on sustainable practices:** Recognizing the growing importance of sustainability, we have dedicated a chapter to green warehousing, outlining strategies for reducing environmental impact and implementing ISO 14000 standards.
3. **Integration of technology:** From warehouse management systems to automation and robotics, this book explores the role of technology in modern warehouse operations, highlighting both benefits and challenges.
4. **Focus on safety and security:** Ensuring the safety and security of warehouse operations is paramount. This book covers occupational safety measures, security protocols, and risk management strategies.
5. **Case studies and best practices:** Real-world examples and lessons from industry leaders provide valuable insights into successful inventory management practices, helping readers apply best practices to their own operations.
6. **Skill development:** This book also addresses the skills and competencies required for efficient warehouse operations, offering guidance on staff training and development.

The concluding chapters summarize key learnings and provide a glimpse into future trends in warehouse management, offering final thoughts on optimizing inventory management.

2. Overview of Warehouse Operations

Warehouses serve as vital parts of the supply chain, bridging the gap not only between raw material sources and factories but also between wholesalers and end consumers. As illustrated in Figure 1, each stage of the supply chain necessitates a storage system to house inventory, underscoring the pivotal role warehouses play in facilitating seamless transitions and ensuring the smooth flow of goods.

2.1 *Why warehouse?*

Warehouses serve as pivotal hubs within supply chain networks, facilitating the storage, handling, and distribution of goods. They act as intermediaries between suppliers and customers, playing a crucial role in ensuring timely delivery and maintaining optimal inventory levels. A well-organized warehouse operation involves various interconnected processes, including receiving, storage, picking, packing, and shipping. Each of these functions

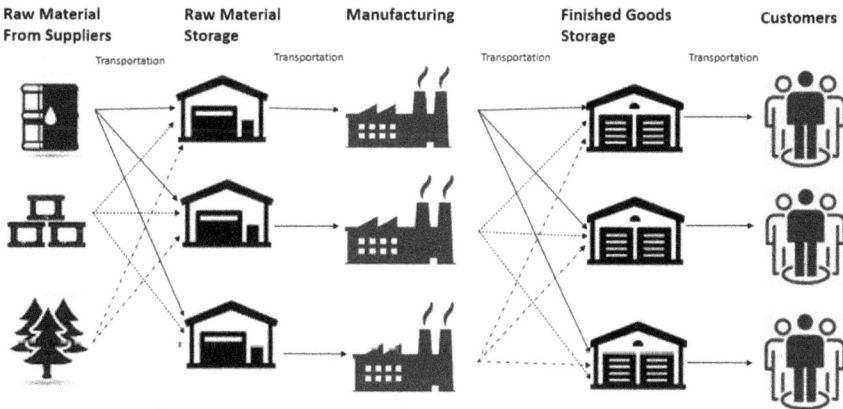

Figure 1. Supply chain storage system.

contributes to the smooth flow of goods through the facility, ultimately impacting customer satisfaction and the bottom line.

Modern warehouse operations are characterized by their complexity and the need for agility to adapt to ever-changing market demands. As technology has improved and new ideas have been used, warehouses have changed into dynamic places where automation with the Internet of Things (IoT), data analytics, and efficient workflows are all important for achieving operational excellence. Hence, the definition of *a warehouse is part of a logistics system that stores products (raw materials/supplies/in process/packaging material/finished goods) at and between the point of origin and the point of consumption, providing information to management on the status, conditions, and disposition of the items being stored.*[1]

3. Warehouse Ownership Decisions

Deciding whether to own or lease warehouse space is a critical strategic decision that affects a company's operational flexibility and financial stability. Firms must weigh the costs and benefits associated with private and public warehousing to determine the most suitable approach.

1. **Cost factors**
    i. **Variable costs:** These are expenses that fluctuate with the level of warehouse activity. For private warehouses, variable costs include labor, utilities, and maintenance. Public warehouses may also charge fees based on the volume of goods stored and handled.
    ii. **Fixed costs:** These are consistent expenses that do not change with the level of activity. In private warehousing, fixed costs encompass rent, insurance, and depreciation of warehouse facilities. Public warehousing typically involves regular fees or contracts that ensure space availability.
2. **Qualitative factors**
    i. **Private warehousing**
       Advantages of private warehousing:
       • In terms of control, the company has a greater degree of control over its products in the warehouse as they have the freedom to perform all activities within the warehouse, such as kitting, bundling, and other value-added activities. Hence, we are able to satisfy customers' requirements without sacrificing customer service.

- In terms of flexibility, a company comes with greater flexibility because it can design layout and storage based on products with special handling or storage requirements. Also, companies can take into consideration their products and customer order patterns in the design stage.
- In terms of cost, over the long run, it would be cheaper by 15–25% to operate if the utilization rate was 75–80% due to economies of scale.
- In terms of human resources, the company can better utilize its resources and cross-train the employees to learn more processes unique to the company. This helps improve productivity by moving employees from a lull activity to a busy one.
- In terms of finance, a company can claim a tax incentive when buying automation equipment. The company can also lower its tax payments by amortizing its property. In Singapore, companies can claim the Productivity and Innovation Credit (PIC) scheme, which provides tax deductions or cash payouts for investments in innovation and productivity improvements.

Disadvantages of private warehousing

- In terms of flexibility, if the company decides to change or remove some SKUs from its product lines, the new SKUs with different products and special handling or storage requirements will affect the design layout, which in turn will affect the smooth flow of operations.
- In terms of finance, it is a long-term commitment, as the company has to put in the initial investment to recoup the benefits. Due to a long payback period or a non-viable return on investment (ROI) hence, many companies put off building their own warehouses.
- In terms of human resources, the company needs to maintain a pool of highly skilled employees to perform warehouse operations. During economic downtime, there are not enough orders to keep the workforce occupied, resulting in lost productivity or wasted resources.

ii. **Public warehousing**
Advantages of public warehousing

- In terms of finance, a company can just lease enough space to run its operations. A company knows its fixed cost of running a warehouse as rented space is fixed every month. Additionally, other fixed costs such as insurance premiums, salaries for permanent staff, and technology subscriptions contribute to a stable financial outlook. Indirect costs such as utilities, maintenance, staff overtime, and packaging materials can be estimated, hence the company is able to estimate P&L even before the end of the month.

- In terms of flexibility, when volume grows, the company can expand to adjacent space next to it or take up bigger space as and when needed to expand the operations. This flexibility allows third-party logistics (3PL) to lease not only space but also material handling equipment from equipment providers based on customer contract duration.

Disadvantages of public warehousing

- In terms of space, securing an ideal warehouse location can prove challenging, especially when considering future expansions. Also, proximity to amenities often inflates rental costs, further complicating matters.
- In terms of finance, normally warehouse owners sign on fixed rates for certain periods (3–5 years) and subsequently, rental rates will increase 7–10% after that. Hence, the company has to make enough profit to cover the increased rate.
- In terms of warehouse design, warehouses featuring pillar-supported layouts may impede operations for companies with specialized handling and storage needs. However, it negates the above problem if the company commits to leasing the whole building for the long term. A landlord could design and build to order facilities to suit specific requirements, mitigating operational hindrances.

4. Optimal Number of Warehouses

Determining the optimal number of warehouses involves balancing various factors to achieve cost efficiency and high service levels. This decision impacts inventory management, transportation costs, and overall supply chain performance:

- **Inventory costs:** More warehouses typically mean higher inventory holding costs, as stock must be distributed across multiple locations.

- **Transportation costs:** Increasing the number of warehouses can reduce transportation costs by shortening delivery distances to customers.
- **Service levels:** More warehouses can improve service levels by enabling faster delivery times and better responsiveness to local market demands.
- **Economies of scale:** Fewer, larger warehouses can achieve economies of scale in operations and inventory management.
- **Risk management:** Distributing inventory across multiple locations can mitigate risks related to disruptions, such as natural disasters or transportation strikes.

5. Centralized vs. Decentralized Warehousing

The choice between centralized and decentralized warehousing depends on the firm's strategic goals, market conditions, and operational requirements.

Centralized warehousing

Centralized warehousing consolidates inventory in a single location or a few strategic locations. This model can lead to significant cost savings and streamlined operations but may also have drawbacks.

- **Advantages:**
  - **Reduction in working capital:** Centralizing inventory reduces the need for excess stock, freeing up capital for other uses.
  - **Lower operational costs:** Consolidated operations can achieve efficiencies in labor, equipment, and facility management.
  - **Reduced inbound and outbound costs:** Fewer locations can simplify logistics and reduce transportation expenses.
  - **Simplified order processing and inventory management:** Centralized systems are easier to manage and monitor.
  - **Better customer service:** A central location can facilitate consistent service standards and quicker response times.
- **Disadvantages:**
  - **High cost of rush delivery:** Centralized locations may require expensive expedited shipping to meet urgent demands.
  - **Slower decision-making processes:** Centralized systems can be less flexible, with longer decision-making chains.

- o **Less flexibility in emergencies:** Centralized warehouses may struggle to adapt quickly to unforeseen disruptions.

Decentralized warehousing

Decentralized warehousing distributes inventory across multiple locations, closer to key markets or customers. This model offers greater flexibility and responsiveness but at higher operational costs.

- **Advantages:**
  - o **Better adaptation to customer needs:** Proximity to customers allows for tailored services and quicker deliveries.
  - o **Faster decision-making:** Localized management can make quicker decisions and respond promptly to changes.
  - o **Better responsiveness in emergencies:** Decentralized systems can adapt more readily to local disruptions or demand spikes.
- **Disadvantages:**
  - o **Higher inbound and outbound costs:** More locations can lead to increased transportation expenses for moving goods between warehouses.
  - o **Additional transportation expenses:** Distributing inventory across multiple sites can result in higher logistics costs.
  - o **Higher operational costs:** Multiple facilities require more resources for staffing, equipment, and maintenance.

6. Decision-Making Hierarchy

Strategic decisions in warehousing are influenced by a hierarchy of considerations, from corporate and business strategies to specific warehousing policies and functional strategies:

- **Corporate and business strategies:** These high-level strategies guide the overall direction and goals of the firm, influencing warehousing decisions.
- **Warehousing policies:** Policies establish the framework for warehousing operations, including guidelines for inventory management, space utilization, and cost control.
- **Functional strategies:** Functional strategies focus on specific areas of warehouse operations, such as logistics, order fulfillment, and customer service.

- **Warehousing decisions:** These are day-to-day decisions regarding warehouse operations, including layout design, staffing, and process improvements.

7. Warehousing Policies

Several factors influence warehousing policies, including industry characteristics, the firm's philosophy, capital availability, product characteristics, economic conditions, competition, seasonality of demand, and the production process:

- **Industry characteristics:** The nature of the industry, including product types and market dynamics, affects warehousing requirements.
- **Firm's philosophy:** A firm's strategic orientation, such as a focus on cost leadership or differentiation, shapes warehousing policies.
- **Capital availability:** The availability of capital influences decisions on warehouse investments and operational expenditures.
- **Product characteristics:** Product size, weight, shelf life, and handling requirements impact warehousing needs and policies.
- **Economic conditions:** Economic factors, such as interest rates and market trends, affect warehousing costs and investment decisions.
- **Competition:** Competitive pressures drive firms to optimize warehousing operations for cost efficiency and service excellence.
- **Seasonality of demand:** Seasonal variations in demand influence warehousing capacity and inventory management strategies.
- **Production process:** The production process, including lead times and batch sizes, affects warehousing needs and policies.

So, what is the function of a warehouse? As shown in Figure 1, warehouses serve a few functions in the supply chain network. The basic needs of a warehouse are as follows:

i. **Raw material/supply storage:** Following extraction, raw materials must undergo storage before processing begins.
ii. **In-process/packaging material storage:** Within manufacturing plants, storage of in-process and packaging materials serves as a buffer to support the production process from raw materials to finished goods.

iii. **Finished goods storage:** Finished goods are stored prior to distribution to wholesalers, retailers, or end-users. They are packaged and await shipping.

iv. **Goods stockpiling:** Stockpiling of goods is undertaken to create a buffer for essential or seasonal products. For instance, the Singapore government stockpiles staple foods such as rice to ensure a consistent supply at reasonable prices for its citizens. The Hallmark card greeting company stocks seasonal greeting cards to prepare for festive season greetings, such as Christmas.

v. **Order fulfillment:** Warehouses facilitate order processing by picking, packing, and shipping products to customers or other distribution points.

Besides storage and fulfillment, the intangible part of the warehouse is to protect the products from loss, theft, and deterioration from natural elements like rain and shine.

Other needs of a warehouse, or especially a modern warehouse, are as follows:

i. *Merge-in-transit integration* enables components of a final product to merge at a designated location before delivery to the customer. For instance, consider a personal computer (PC) manufacturer producing CPU units in Country A and monitor units in Country B. If a customer in Country C orders a PC, the components from Countries A and B are shipped to Country C for integration before delivery to the customer. Additionally, to accommodate language preferences and comply with local government regulations, the final assembly may be postponed. This could involve customizing country-specific user manuals, electrical plugs, and product license labels in the destination country.

ii. *Consolidation* facilitates the bundling of products to achieve full container or truck loading, thereby capitalizing on the cost efficiencies offered by transportation economies of scale. Additionally, it enables companies to offer diverse product assortments to customers at a single location rather than dispatching them from multiple warehouses. See Figure 2 for more details.

iii. *Breakbulk facilities* allow warehouses to receive large shipments from manufacturing plants in FCL (Full Container Load) for further breakdown to smaller quantities to customers in LCL (Loose Container Load) who normally need smaller quantities.

Consolidation and Breakbulk Illustration

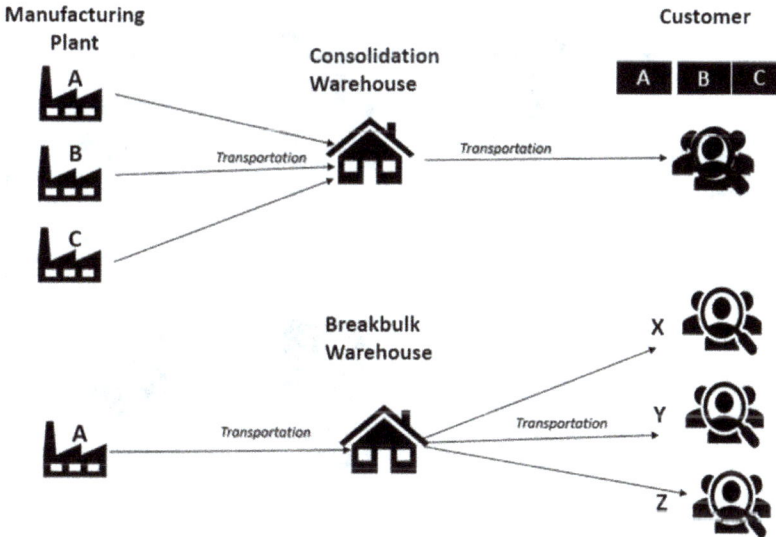

Figure 2.   Consolidation and breakbulk illustration.

iv. *Spot stocks* specifically target stocks, particularly those associated with manufacturing limited and highly seasonal product lines. A portion of the firm's product inventory is strategically positioned, or "spot stocked," within warehouses to swiftly fulfill customer orders during crucial marketing seasons. This tactic allows inventory to be placed in various markets near key customers, right before peak seasonal sales. Particularly advantageous for e-commerce enterprises aiming to highlight specific products during targeted sales campaigns.

v. *Cross-docking operations* facilitate the direct transfer of incoming goods to outbound transportation vehicles, thereby reducing the expenses associated with handling and storage. Larger warehouses or distribution centers typically deploy this to consolidate incoming goods and storage, maximizing the container or truck load, as shown in Figure 3.

vi. *Returns management* facilitates warehouses handling product returns and managing reverse logistics processes, including inspection, sorting, and disposition of returned items.

Cross-Docking Illustration

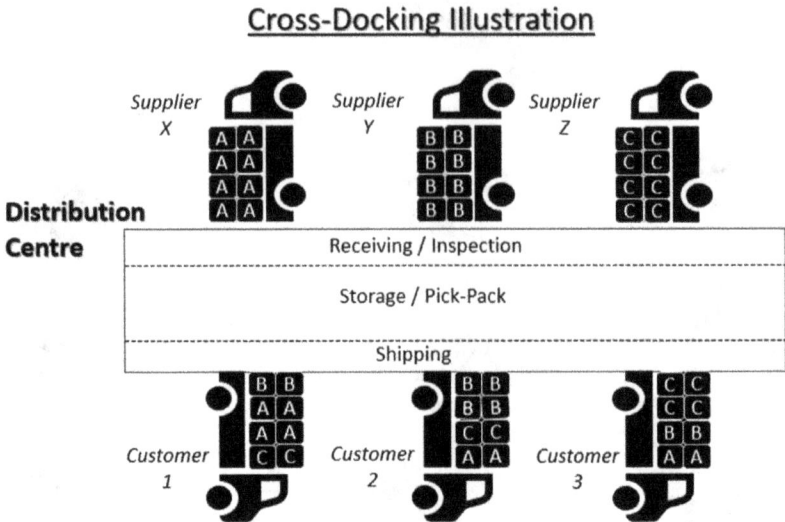

Figure 3.   Cross-docking illustration.

8. Importance of Effective Inventory Management

Effective inventory management lies at the heart of successful warehouse operations. It entails maintaining the right balance of stock levels to meet customer demand while minimizing carrying costs and avoiding stockouts or overstock situations.

Optimizing inventory management offers several compelling benefits, including the following:

- **Cost reduction:** Efficient inventory management helps minimize holding costs associated with storing excess inventory and mitigates the risk of obsolescence. By streamlining inventory levels, organizations can free up valuable capital and improve cash flow.
- **Enhanced customer satisfaction:** Timely order fulfillment and accurate inventory levels contribute to a positive customer experience. By ensuring products are readily available when needed, businesses can strengthen customer loyalty and foster long-term relationships.
- **Operational efficiency:** Well-managed inventory enables smoother warehouse operations, reducing the likelihood of bottlenecks and

delays. By implementing effective inventory control measures, organizations can optimize space utilization, reduce handling times, and enhance overall productivity.

- **Data-driven insights:** Inventory management systems provide valuable data insights that enable informed decision-making. By analyzing trends, demand patterns, and inventory turnover rates, organizations can anticipate market fluctuations, identify opportunities for improvement, and optimize supply chain performance.

In today's competitive business landscape, effective inventory management is not just a necessity but a strategic imperative. By prioritizing inventory optimization and adopting best practices, warehouse managers can position their operations for sustained growth and profitability in the long run. Lastly, in addition to discussing the type of warehouse, it's crucial to consider the operating environment of the warehouse. For example, running a warehouse for FMCG (Fast-Moving Consumer Goods) is vastly different from one for chemical DG (Dangerous Goods). Let's explore the specifics of chemical warehouses in more detail, as this topic is often overlooked in most literature.

In Singapore, constructing a chemical warehouse requires adherence to stringent fire safety regulations to protect people, property, and the environment. The key requirements are outlined in the Fire Code and other relevant guidelines issued by the SCDF (Singapore Civil Defense Force). Here are some major points:

1. **Fire safety certificate (FSC):** Before construction, you must obtain a Fire Safety Certificate from SCDF by submitting fire safety plans and designs for approval.
2. **Hazardous materials (HazMat) regulations:** The storage of hazardous materials is regulated by SCDF. A Hazardous Materials License is required if your warehouse will store regulated quantities of hazardous substances.
3. **Fire protection systems:** Chemical warehouses must be equipped with appropriate fire protection systems, including
   - automatic fire sprinklers,
   - smoke detection Systems,
   - fire extinguishers,
   - fire hose reels and hydrants.

4. **Compartmentalization:** To prevent the spread of fire, warehouses should be partitioned with fire-resistant walls and floors.
5. **Proper ventilation:** Proper ventilation is required to prevent the accumulation of flammable vapors.
6. **Emergency exits and escape routes:** Adequate emergency exits and clear escape routes must be provided, clearly marked, and unobstructed at all times.
7. **Separation distances:** Specific requirements exist for separation distances between the warehouse and other buildings or boundaries, especially if storing flammable or explosive materials.
8. **Spill containment:** To prevent environmental contamination, facilities for containing and managing chemical spills must be put in place.
9. **Emergency response plan:** An emergency response plan tailored to the specific chemicals stored must be prepared and regularly updated. Staff should be trained in emergency procedures.
10. **Regular inspections and maintenance:** To ensure that fire safety systems and equipment are operational, regular inspections and maintenance are mandatory.
11. **Compliance with standards:** Ensure compliance with relevant Singapore Standards (SS) and Codes of Practice, such as SS 532 for the storage of flammable liquids.

Also, in Singapore, different government agencies are in charge of different classes of dangerous goods. In general, explosives, flammables, oxidizing agents, and others (class 1, 2, 3, 4, 5, 8, 9) are under the jurisdiction of the SCDF agency. However, the NEA (National Environment Agency) takes charge of toxic and infectious substances (class 6). Whereas HSA (Health Safety Authority) controls radioactive materials (class 7). You can refer to the classification from this link: https://www.caas.gov.sg/docs/default-source/docs---srg/fs/dg/annex---classification-of-dangerous-goods.pdf.

To effectively contain any spillages, a Dangerous Goods (DG) warehouse typically constructs its racking 1.0–3.0 meters below the loading bays. Additionally, a containment area measuring 3.0 × 3.0 cubic meters should be built beneath the racking floor. This containment storage will collect DG liquids, which will be directed there with the use of spill kit materials in the event of a spillage.

We hope that this book serves as a valuable resource for warehouse owners, managers, and professionals seeking to enhance their understanding and management of warehouse inventory. By applying the principles and practices outlined in this book, we believe that readers can achieve significant improvements in their warehouse operations, ultimately leading to greater efficiency, reduced costs, and enhanced customer satisfaction.

Reference

Council of Supply Chain Management Professionals. (n.d.). *Warehousing operations*. Retrieved December 7, 2024, from https://cscmp.org/CSCMP/Certify/Fundamentals/Warehousing_Operations.aspx.

Chapter 2

Warehouse Layout Design

Learning Outcome

By the end of this chapter, you should be able to do the following:
1. Explain the key principles of warehouse design.
2. Identify the type of warehouse layout.
3. Recognize factors influencing warehouse layout decisions.
4. Describe best practices in warehouse layout.

1. Introduction

Warehouse layout design is a critical aspect of warehouse operations that directly impacts supply chain management's efficiency, productivity, and overall performance. A well-designed warehouse layout optimizes space utilization, improves material handling, enhances safety, and facilitates better inventory management. This chapter delves into the fundamental principles of warehouse layout design, exploring different layout types, factors influencing layout decisions, and best practices for creating an efficient warehouse environment.

2. Key Principles of Warehouse Layout Design

- **Space utilization:** Effective use of available space is paramount in warehouse layout design. Vertical space should be maximized using tall storage racks and shelving units, while horizontal space should be organized to minimize wasted areas. Frazelle (2016) emphasized the importance of optimizing storage density to reduce space waste and improve operational efficiency.

- **Flow of goods:** Ensuring a smooth and logical flow of goods is essential for efficient warehouse operations. This involves designing the layout to facilitate easy movement of products from receiving to storage, picking, packing, and shipping. Richards (2017) highlighted the significance of minimizing travel distances and avoiding bottlenecks in the flow of goods.
- **Accessibility:** All products should be easily accessible to warehouse staff. This can be achieved by strategically placing high-turnover items and utilizing appropriate storage methods such as pallet racks and bin shelving. Bartholdi and Hackman (2019) discussed the importance of accessibility in reducing pick times and enhancing operational speed.
- **Flexibility:** A flexible warehouse layout can adapt to changing business needs and inventory levels. Modular shelving units, adjustable racking systems, and open floor plans can accommodate fluctuations in product volume and variety. Garza-Reyes *et al.* (2014) underscored the need for adaptability in warehouse layouts to respond to market dynamics and seasonal demands.
- **Safety:** Ensuring the safety of warehouse staff and goods is a crucial aspect of layout design. This involves clear aisle markings, adequate lighting, proper ventilation, and compliance with safety regulations. Fichtinger *et al.* (2015) advocated for incorporating safety measures into warehouse layout planning to prevent accidents and improve working conditions.

3. Types of Warehouse Layouts

Figure 1 displays each of the warehouse layouts.

3.1 *U-shaped layout*

In a U-shaped layout, the receiving and shipping docks are strategically located on the same side of the building. This design ensures that the storage and picking areas are situated in between the docks. By arranging the

U-SHAPED WAREHOUSE   **L-SHAPED WAREHOUSE**   **I-SHAPED WAREHOUSE**

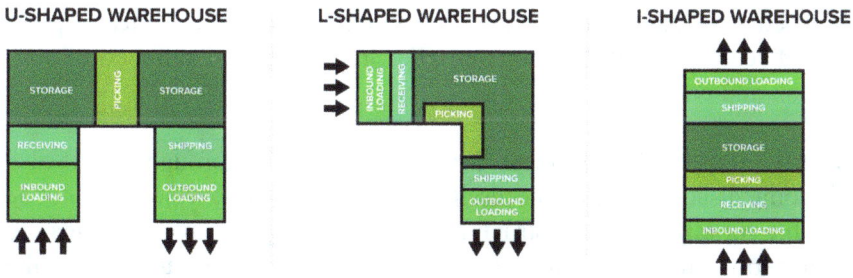

Figure 1. Types of warehouse layouts.

warehouse in this manner, the travel distance for goods is minimized, thereby facilitating efficient material handling.

Advantages: The U-shaped layout offers several benefits. First, it improves the flow of goods within the warehouse. Second, it allows for easy supervision of operations, as all activities are concentrated in a central area. Lastly, it promotes the efficient use of space, as the layout ensures that all available areas are utilized effectively.

Disadvantages: However, the U-shaped layout also has some drawbacks. It may require more floor space compared to other layouts, which could be a limiting factor for warehouses with restricted space. Additionally, this layout can be less flexible for future expansion, as the U-shape confines the areas where new sections can be added.

3.2 *Straight-through (I-shaped) layout*

A straight-through, or I-shaped, layout consists of receiving docks on one end of the building and shipping docks on the other. The storage areas are positioned between these two docks, creating a linear flow of goods from one end of the warehouse to the other. This design supports a streamlined flow of goods, thereby reducing handling times and improving overall efficiency.

Advantages: The straight-through layout provides several advantages. It ensures a streamlined flow of goods through the warehouse, minimizing the need for excessive handling. This layout also promotes a clear separation of receiving and shipping activities, reducing the likelihood of congestion and confusion.

Disadvantages: Despite these benefits, the straight-through layout also has some disadvantages. It necessitates larger buildings, which may not be feasible for all warehouse structures. Additionally, this layout can lead to higher travel distances for staff, as employees may need to traverse the length of the warehouse to move between receiving and shipping areas.

3.3 *L-shaped layout*

In an L-shaped layout, the receiving and shipping docks are located on adjacent sides of the building, with the storage areas forming the "L" shape. This layout combines elements of both U-shaped and straight-through designs, offering a flexible and efficient use of space.

Advantages: The L-shaped layout provides several benefits. Its flexible design allows for efficient use of available space, accommodating various storage and picking needs. The layout also promotes a good flow of goods, ensuring that materials move smoothly through the warehouse.

Disadvantages: However, the L-shaped layout can be complex to manage. It requires careful planning to avoid congestion and ensure that all areas of the warehouse are utilized effectively. Additionally, the layout may require more intricate supervision to maintain an efficient flow of goods. Table 1. lists a comparison of each warehouse layout.

4. Factors Influencing Warehouse Layout Decisions

- **The nature of goods:** The physical characteristics of the goods, such as size, weight, and perishability, influence the choice of storage systems and layout design. For example, bulky items may require wide aisles and heavy-duty racking, while perishable goods need refrigerated storage areas. Al-Refaie *et al.* (2020) discussed the impact of product characteristics on warehouse layout and storage solutions.

Table 1.   Comparison of the three warehouse layouts based on their characteristics.

Factor	U-Shaped Layout	I-Shaped Layout	L-Shaped Layout
Description	Receiving and shipping docks on the same side with storage in between	Receiving docks on one end and shipping docks on the opposite end	Receiving and shipping docks on adjacent sides with storage forming an "L" shape
Flow of Goods	Improved flow due to centralized operations	Streamlined linear flow from receiving to shipping	Good flow by combining U-shaped and I-shaped elements
Supervision	Easy supervision with centralized operations	Clear separation of receiving and shipping activities	Requires more intricate supervision to avoid congestion
Space Utilization	Efficient use of space but may require more floor space	Efficient but needs longer buildings	Flexible and efficient use of space
Handling Times	Minimized travel distance for goods	Reduced handling times due to linear flow	Efficient handling but requires careful planning
Flexibility for Expansion	Less flexible for expansion due to the confined U-shape	Limited flexibility due to the need for longer buildings	Flexible design accommodating various needs
Complexity	Simple to manage with concentrated areas	Simple to manage with clear separation	More complex to manage and plan
Travel Distance for Staff	Moderate travel distance for staff	Higher travel distances for staff	Moderate travel distance but requires planning to avoid congestion
Example	Suitable for warehouses needing centralized operations and easy supervision	Ideal for large, elongated warehouses with clear separation of activities	Best for warehouses needing flexible and efficient space utilization

- **Volume of goods:** Inventory volume and turnover rate determine storage capacity and layout requirements. High-volume warehouses need efficient space utilization and robust storage systems to handle large quantities of goods. Ahmadi *et al.* (2017) highlighted the need for scalable storage solutions to accommodate varying inventory levels.
- **Operational workflow:** The workflow processes, including receiving, storage, picking, packing, and shipping, dictate the layout design. An optimized workflow minimizes handling times and improves overall efficiency. Ding and Kaminsky (2020) emphasized the importance of aligning warehouse layout with operational workflows to enhance productivity.
- **Technology integration:** Incorporating technology such as automated storage and retrieval systems (AS/RS), conveyor belts, and warehouse management systems (WMS) can influence layout design. These technologies improve accuracy, speed, and efficiency in warehouse operations. Haleem and Javaid (2019) discussed the role of technology in modern warehouse design and operations.
- **Future growth:** In warehouse layout design, planning for future expansion and scalability is crucial. Flexible layouts that allow for easy reconfiguration and expansion can accommodate business growth and changing needs. Garza-Reyes (2015) advocated for designing warehouses with future scalability in mind to support long-term business growth.

5. Best Practicies for Warehouse Layout Design

- **Conduct a thorough analysis:** A comprehensive analysis of current operations, inventory levels, and future needs is essential before designing the warehouse layout. This includes evaluating space requirements, material handling equipment, and workflow processes. Simchi-Levi *et al.* (2008) recommended conducting detailed operational analysis to inform layout design decisions.
- **Incorporate lean principles:** Applying lean principles to warehouse layout design can reduce waste, improve efficiency, and enhance productivity. This involves eliminating unnecessary steps, optimizing space utilization, and streamlining processes. Garza-Reyes (2015) highlighted the benefits of lean principles in creating efficient warehouse layouts.

- **Engage stakeholders:** Involving key stakeholders in the layout design process, including warehouse staff, management, and supply chain partners, ensures that all perspectives are considered and the design meets operational needs. Mentzer *et al.* (2007) emphasized the importance of stakeholder engagement in successful warehouse layout planning.
- **Utilize simulation tools:** Simulation tools can model different layout scenarios and assess their impact on warehouse operations. This helps identify potential issues, optimize layout design, and improve decision-making. Bartholdi and Hackman (2019) advocated for the use of simulation tools in warehouse layout design to enhance accuracy and efficiency.
- **Prioritize safety and ergonomics:** Safety and ergonomic design reduces the risk of accidents and improves working conditions for warehouse staff. This includes proper aisle widths, adequate lighting, and ergonomic workstations. Fichtinger *et al.* (2015) highlighted the importance of incorporating safety and ergonomics into warehouse layout design.

6. Case Study: QuickMart Warehouse Layout Design

Background: A fast-moving consumer goods (FMCG) company, "QuickMart," is expanding its operations and needs a new warehouse layout to handle increased volumes of products. The company deals with a variety of products, including food items, beverages, household products, and personal care items. The warehouse will serve as a central hub for receiving goods from suppliers, storing them, and shipping them to various retail locations.

Objectives: The primary objectives for designing QuickMart's new warehouse layout are to optimize space utilization, ensure smooth and efficient movement of goods, improve safety and ergonomics, and allow for future expansion and adaptability to changing product lines and volumes.

Warehouse specifications: The total area of the warehouse is 50,000 square feet, with 10 dock doors available: 5 designated for receiving and 5 for shipping. The warehouse will use a combination of pallet racks, shelving, and bulk storage. Additionally, there will be temperature-controlled zones for perishable items.

7. Selected Layout: Straight-Through Layout

The straight-through layout is selected for QuickMart's new warehouse due to its clear separation of receiving and shipping areas, which helps reduce congestion and streamline the flow of goods (see Figure 2). This layout is suitable for handling high volumes and allows for efficient processing of diverse product categories.

Layout Design:

1. **Receiving area:** The receiving area is located at one end of the warehouse, with five dock doors dedicated to receiving. This area contains

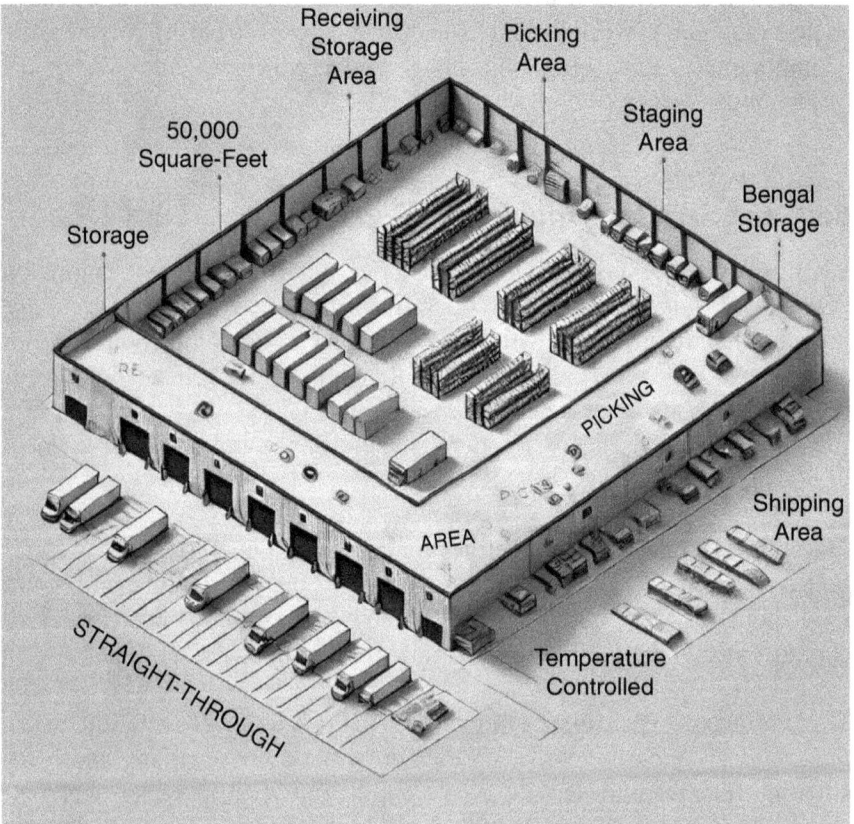

Figure 2.   Proposed layout for the new warehouse layout.

inspection and staging zones for incoming goods, as well as separate zones for perishable and non-perishable items.

2. **Storage area:** The central part of the warehouse is designated as the storage area. This area features pallet racks for bulk storage of high-volume items, shelving units for smaller, slower-moving items, and temperature-controlled zones for perishable goods. Clear aisles are maintained for forklift and pallet jack movement.

3. **Picking area:** The picking area is adjacent to the storage area, where designated zones are set up for order picking and packing. Ergonomically designed workstations are provided to reduce manual handling, and automated picking systems are installed for high-volume items.

4. **Shipping area:** The shipping area is located at the opposite end of the warehouse from the receiving area. It includes five dock doors dedicated to shipping, consolidation, and staging areas for outbound orders, with separate zones for different shipping methods, such as palletized and parcel.

5. **Support areas:** The warehouse also includes support areas, such as office space for administrative tasks, break rooms and restrooms for staff, and a maintenance area for equipment.

Implementation Steps:

1. **Initial setup:** The first step is to conduct a detailed analysis of product types and volumes. Following this, specific zones for receiving, storage, picking, and shipping will be designated. Pallet racks, shelving, and temperature-controlled storage will be installed according to the design.

2. **Operational workflow:** Standard operating procedures (SOPs) for receiving, storage, picking, and shipping will be developed. Staff will be trained on new processes and safety protocols to ensure smooth operations.

3. **Technology integration:** A Warehouse Management System (WMS) will be implemented to track inventory and manage operations. Automated picking systems will be installed for high-volume items to enhance efficiency.

4. **Continuous improvement:** The warehouse performance will be monitored regularly, and adjustments will be made as needed. Regular safety audits and staff training sessions will be conducted to maintain high standards of safety and efficiency.

8. Summary

Designing a warehouse layout for QuickMart involves optimizing space utilization, enhancing the flow of goods, improving safety, and ensuring scalability. The straight-through layout provides a clear separation of receiving and shipping areas, reducing congestion and improving efficiency. By implementing this layout and following the outlined steps, QuickMart can achieve a highly efficient and flexible warehouse operation that meets its growing needs. Warehouse layout design is a critical factor in achieving efficient and effective warehouse operations. By considering key principles, understanding different layout types, and incorporating best practices, businesses can create warehouse environments that optimize space utilization, improve workflow, enhance safety, and support long-term growth. As technology continues to evolve, future warehouse layouts will increasingly integrate advanced systems and automation to further streamline operations and meet the demands of a dynamic supply chain landscape.

References

Ahmadi, H.B., Kusi-Sarpong, S., and Rezaei, J. (2017). Assessing the social sustainability of supply chains using the Best Worst Method. *Resources, Conservation, and Recycling*, 126, 99–106.

Al-Refaie, A., Al-Tahat, M., and Lepkova, N. (2020). Modelling relationships between agility, lean, resilient, green practices in cold supply chains using ISM approach. *Technological & Economic Development of Economy*, 26(4), 675–694.

Bartholdi, J.J. and Hackman, S.T. (2019). *Warehouse & Distribution Science*, 1st edn. The Supply Chain and Logistics Institute, Atlanta: Georgia Institute of Technology.

Ding, S. and Kaminsky, P.M. (2020). Centralized and decentralized warehouse logistics collaboration. *Manufacturing & Service Operations Management*, 22(4), 812–831.

Fichtinger, J., Ries, J.M., Grosse, E.H., and Baker, P. (2015). Assessing the environmental impact of integrated inventory and warehouse management. *International Journal of Production Economics*, 170, 717–729.

Frazelle, E.H. (2016). *World-Class Warehousing and Material Handling*, 2nd ed. New York: McGraw-Hill Education.

Garza-Reyes, J.A. (2015). Lean and green — A systematic review of the state of the art literature. *Journal of Cleaner Production*, 102, 18–29.

Garza-Reyes, J.A., Winck Jacques, G., Lim, M.K., Kumar, V., and Rocha-Lona, L. (2014). Lean and green–synergies, differences, limitations, and the need for Six Sigma. In: Grabot, B., Vallespir, B., Gomes, S., Bouras, A., Kiritsis, D. (eds). *Advances in Production Management Systems. Innovative and Knowledge-Based Production Management in a Global-Local World.* APMS 2014. IFIP Advances in Information and Communication Technology, Vol. 439. Berlin, Heidelberg: Springer.

Haleem, A. and Javaid, M. (2019). Additive manufacturing applications in Industry 4.0: A Review. *Journal of Industrial Integration and Management*, 4(04), 1930001.

Mentzer, J.T., Myers, M.B., and Stank, T.P. (2007). *The Global Supply Chain Management Handbook*, 1st edn. USA: Sage Publications.

Richards, G. (2017). *Warehouse Management: A Complete Guide to Improving Efficiency and Minimizing Costs in the Modern Warehouse*, 1st edn. USA: Kogan Page Publishers.

Simchi-Levi, D., Kaminsky, P., and Simchi-Levi, E. (2008). *Designing and Managing the Supply Chain: Concepts, Strategies, and Case Studies*, 1st edn. USA: McGraw-Hill.

Chapter 3

Inventory Classification and Segmentation

Learning Outcome

By the end of this topic, you should be able to do the following:

1. Identify the ABC analysis: Prioritizing inventory.
2. Interpret the XYZ analysis: Demand forecasting.
3. Describe the SKU rationalization and standardization.

1. Introduction

This chapter delves into the crucial role of a fundamental aspect of efficient inventory management. As businesses strive for optimization in their operations, understanding how to effectively categorize and prioritize inventory becomes paramount. This chapter aims to equip learners with the necessary skills to steer this ground accurately. First, this chapter explores the ABC analysis, a methodical approach to prioritizing inventory based on its importance and value. By mastering this analysis, learners will learn how to allocate resources wisely, focusing efforts where they yield the most significant returns. Moving forward, this chapter unravels the complexities of the XYZ analysis, which extends beyond mere prioritization to forecast demand. Through this analysis, learners will gain insights into predicting future inventory needs, allowing for proactive decision-making and minimizing stockouts or excess inventory. Finally, this chapter delves into SKU rationalization and standardization, shedding light on the process of streamlining product offerings to enhance efficiency and reduce complexity. Understanding how to optimize SKU assortments can lead to improved inventory turnover and better resource utilization. By the end of this chapter, learners will emerge with a comprehensive understanding of inventory classification and segmentation methodologies, equipped to make informed decisions that drive efficiency and profitability within the organization.

2. ABC Analysis: Prioritizing Inventory

2.1 *ABC analysis overview*

ABC analysis is a method used in inventory management to categorize items based on their significance and value to the business, helping prioritize resource allocation and management efforts. By applying ABC analysis, businesses can focus their efforts on managing A-items closely to maximize revenue, ensure B-items are adequately stocked, and optimize C-items to reduce carrying costs and improve overall efficiency. This strategic approach enables better inventory control, resource allocation, and financial performance. Items are divided into three categories:

1. **A-items:** These are high-value items that, although few in number, contribute the most to the company's revenue. For example, a luxury car manufacturer might classify its top-of-the-line models as A-items. Despite being a small percentage of total inventory, these models generate substantial revenue and require careful management to avoid stockouts.
2. **B-items:** These items are of moderate value and account for an intermediate portion of the inventory. They require regular but less intense attention compared to A-items. For instance, a mid-range smartphone in an electronics store could be considered a B-item. These products are sold more frequently than high-end items but contribute less to overall revenue per unit.
3. **C-items:** These are low-value items that, although numerous, contribute the least to the company's revenue. An example would be accessories such as phone cases or chargers in the same electronics store.

While they represent a large portion of the inventory, their individual impact on revenue is minimal.

2.2 *ABC analysis steps*

ABC analysis involves several systematic steps to categorize inventory items based on their importance and value to the business. By following the below steps, businesses can effectively prioritize their inventory management efforts, ensuring optimal resource allocation and improving operational efficiency. Key steps involved are as follows:

1. **Data collection:** Gather comprehensive data on all inventory items, including their annual usage or sales volume, unit cost, and total value. This information forms the basis for the analysis.
2. **Calculate the annual usage value:** For each item, calculate the annual usage value by multiplying the annual usage quantity by the unit cost. This helps determine the total monetary contribution of each item.
3. **Sort items by value:** Arrange the items in descending order based on their annual usage value. This ranking helps identify the items that contribute the most to the overall inventory value.
4. **Calculate cumulative totals and percentages:** Calculate the cumulative total of the annual usage values and the cumulative percentage of these totals relative to the overall inventory value. This step aids in visualizing the distribution of inventory value across different items.
5. **Classify items into A, B, and C categories:**
   - **A-items:** Typically, the top 10–20% of items that account for around 70-80% of the total inventory value.
   - **B-items:** The next 20–30% of items that contribute about 15–25% of the total inventory value.
   - **C-items:** The remaining 50–70% of items that constitute only 5–10% of the total inventory value.

2.3 *ABC analysis example*

Table 1 shows the annual usage or sales volume, unit cost, and total annual usage value for each inventory item.

Table 1.   Annual usage of each inventory item.

Item	Annual Usage (units)	Unit Cost ($)	Total Annual Usage Value ($)
High-End Smartphone	500	800	400,000
Mid-Range Laptop	300	600	180,000
Budget Earbuds	2,000	20	40,000
Phone Charger	1,000	10	10,000
Smartwatch	400	150	60,000

To proceed with ABC analysis based on the collected data, follow these four steps:

Step 1: Calculate the Total Annual Usage Value.

Item	Total Annual Usage Value ($)
High-End Smartphone	400,000
Mid-Range Laptop	180,000
Budget Earbuds	40,000
Phone Charger	10,000
Smartwatch	60,000

Step 2: Sort Items by Total Annual Usage Value.

Item	Total Annual Usage Value ($)
High-End Smartphone	400,000
Mid-Range Laptop	180,000
Smartwatch	60,000
Budget Earbuds	40,000
Phone Charger	10,000

Step 3: Calculate Cumulative Totals and Percentages.

Item	Total Annual Usage Value ($)	Cumulative Total ($)	Cumulative Percentage (%)
High-End Smartphone	400,000	400,000	(400,000/690,000) × 100 = 57.97
Mid-Range Laptop	180,000	580,000	(580,000/690,000) × 100 = 84.06
Smartwatch	60,000	640,000	(640,000/690,000) × 100 = 92.75
Budget Earbuds	40,000	680,000	(680,000/690,000) × 100 = 98.55
Phone Charger	10,000	690,000	(690,000/690,000) × 100 = 100

Step 4: Classify Items into A, B, and C Categories.

Category	Item	Cumulative Percentage (%)
A	High-End Smartphone	57.97
A	Mid-Range Laptop	84.06
B	Smartwatch	92.75
C	Budget Earbuds	98.55
C	Phone Charger	1000

This classification method helps in prioritizing inventory management efforts. High-end smartphones and mid-range laptops (A-items) need close monitoring and tighter control. Smartwatches (B-items) require regular review, while budget earbuds and phone chargers (C-items) can be managed with more relaxed controls.

2.4 *ABC analysis advantages*

ABC analysis offers several advantages in inventory management, making it a valuable tool for businesses aiming to optimize their operations.

ABC analysis offers a systematic approach to inventory management, enabling businesses to focus on high-value items, optimize stock levels, reduce costs, and improve overall efficiency and profitability. Some key benefits include the following:

1. **Prioritization of resources:** By categorizing inventory items into A, B, and C groups based on their value and significance, businesses can allocate their resources more effectively. A-items, being the most valuable, receive more attention and tighter control, ensuring their availability and minimizing stockouts.
2. **Improved inventory control:** ABC analysis helps in maintaining optimal stock levels. Businesses can focus on A-items to ensure they are always in stock, while C-items, which are less critical, can be ordered in smaller quantities or on demand, reducing carrying costs.
3. **Enhanced forecasting and planning:** Understanding the demand patterns and value contribution of different inventory items aids in more accurate demand forecasting and better planning. This ensures that inventory levels align with actual sales trends, reducing excess inventory and obsolescence.
4. **Cost reduction:** By focusing on high-value items and optimizing stock levels for lower-value items, businesses can reduce storage and handling costs. This leads to overall cost savings and improved profitability.
5. **Increased efficiency:** With a clear understanding of which items are most important, businesses can streamline their inventory management processes. This increases operational efficiency by reducing time spent on less critical items and focusing efforts on managing high-impact items effectively.
6. **Better supplier management:** ABC analysis helps businesses negotiate better terms with suppliers for high-value A-items, as these items are crucial to the business. Improved supplier relationships can lead to better pricing, discounts, and more reliable supply chains.
7. **Informed decision-making:** The insights gained from ABC analysis provide a data-driven basis for making inventory-related decisions. This reduces reliance on intuition or guesswork, leading to more rational and effective inventory management strategies.
8. **Focus on profitability:** By identifying and prioritizing items that contribute most to the company's revenue, businesses can focus on

enhancing the profitability of these items. This strategic approach ensures that the efforts align with the overall financial goals of the organization.

2.5 *ABC analysis disadvantages*

These disadvantages highlight the need for businesses to complement ABC analysis with other inventory management techniques and regularly review and update their classifications to ensure optimal inventory control and decision-making. Five main disadvantages of ABC analysis are as follows:

1. **Simplicity and oversimplification:** ABC analysis is based on a straightforward classification system that may oversimplify complex inventory situations. It primarily considers the monetary value of items, potentially overlooking other critical factors, such as lead time, item criticality, demand variability, and strategic importance.
2. **Static nature:** The classification in ABC analysis often relies on historical data, which may not account for future changes in demand or market conditions. This can result in outdated or inaccurate classifications if not regularly updated, leading to suboptimal inventory management decisions.
3. **Subjectivity in categorization:** The thresholds for categorizing items into A, B, and C groups can be subjective and arbitrary. Different businesses may set different thresholds, leading to inconsistencies and potentially suboptimal classifications. This subjectivity can affect the reliability and effectiveness of the analysis.
4. **Potential neglect of lower-value items:** By concentrating on high-value A-items, businesses might neglect B and C-items, which can also be crucial for operations. For instance, a low-cost component (C-item) might be essential for producing a high-value product. Neglecting these items can disrupt production and affect overall operational efficiency.
5. **Maintenance effort and resource requirements:** Regular updates and reviews are necessary to keep the ABC classification relevant and accurate. This ongoing maintenance can be time-consuming and require dedicated resources. Failure to maintain the analysis can lead to outdated classifications and ineffective inventory management.

Self Check 1

1. How does the simplicity and potential oversimplification of ABC analysis affect the accuracy of inventory management?
2. What challenges arise from the need for regular updates and maintenance in ABC analysis?

---

3. XYZ Analysis: Demand Forecasting

3.1 *XYZ analysis overview*

XYZ analysis is a method used in inventory management to classify items based on the variability and predictability of their demand. This analysis helps businesses understand how demand fluctuates and allows for better forecasting and inventory control. By classifying inventory items into X, Y, and Z categories, businesses can better manage stock levels, optimize ordering processes, and improve overall inventory efficiency.

The items are divided into three categories:

1. **X-items:** These items have very consistent and predictable demand with minimal variability. For example, a daily-use product such as milk in a grocery store might be classified as an X-item, as its demand remains steady over time. For example, bottled water has a stable and consistent demand throughout the year. It rarely changes, making it an X-item.
2. **Y-items:** These items have moderate variability in demand. They are somewhat predictable but can have seasonal or cyclical fluctuations. An example would be holiday decorations, which see a spike in demand during specific times of the year but are otherwise stable. For instance, winter coats experience moderate demand variability. They are in high demand during the winter season but lower during other months, classifying them as Y-items.
3. **Z-items:** These items have highly variable and unpredictable demand, making them challenging to forecast. An example could be emergency repair parts, which are only needed sporadically and without a clear pattern. For example, fashion accessories such as trendy

handbags have unpredictable demand, influenced by fashion trends and seasons, making them Z-items.

3.2 *ABC analysis vs. XYZ analysis*

ABC analysis and XYZ analysis are both inventory management techniques used to categorize items, but they focus on different aspects of inventory control. Combining ABC and XYZ analyses can provide a more comprehensive approach to inventory management. For example, a high-value item with unpredictable demand (A–Z) requires different strategies compared to a low-value item with consistent demand (C–X). This combined approach helps in creating tailored strategies for different inventory segments, optimizing both cost and service levels.

	ABC Analysis	XYZ Analysis
Focus	Prioritizes inventory based on monetary value and contribution to overall revenue.	Categorizes inventory based on demand variability and predictability.
Classification Criteria	A-items: High-value items with the highest contribution to revenue. B-items: Moderate-value items with intermediate revenue contribution. C-items: Low-value items with the least contribution to revenue.	X-items: Items with consistent, predictable demand and low variability. Y-items: Items with moderate demand variability and some level of predictability, often seasonal or cyclical. Z-items: Items with highly variable and unpredictable demand.
Purpose	Helps in prioritizing management efforts, focusing on high-value items for better control and optimization.	Aids in demand forecasting and inventory control by understanding the variability of demand.
Example	In an electronics store: A-items: High-end smartphones. B-items: Mid-range laptops. C-items: Phone chargers.	In a retail store: X-items: Bottled water (steady demand). Y-items: Winter coats (seasonal demand). Z-items: Trendy fashion accessories (unpredictable demand).

	ABC Analysis	XYZ Analysis
Key Differences	Primary Criteria: Focuses on the value and revenue contribution of items.	Primary Criteria: Focuses on demand variability and predictability.
	Management Approach: Prioritizes resource allocation and control efforts based on item value.	Management Approach: Focuses on demand forecasting and planning based on demand patterns.
	Application: Useful for financial prioritization and cost management.	Application: Useful for demand forecasting, planning, and managing stock levels.

3.3 *XYZ analysis steps*

Conducting XYZ analysis involves a systematic approach to classify inventory items based on their demand variability and predictability. By following these steps, businesses can effectively conduct XYZ analysis to improve their inventory management practices and better handle the variability in demand for different items.

Step 1: Data Collection
- **Gather historical data:** Collect data on the demand for each inventory item over a specific period (e.g. monthly sales data for the past year).
- **Calculate demand:** Determine the quantity sold or used for each item during the period.

Step 2: Calculate the Coefficient of Variation (CV)
- **Mean demand:** Calculate the average demand for each item.
- **Standard deviation:** Calculate the standard deviation of demand for each item.
- **Coefficient of variation (CV):** Calculate CV using the formula $CV = \frac{\text{Standard Deviation}}{\text{Mean Demand}}$.

Step 3: Classify Items into X, Y, and Z Categories
- **Set thresholds:** Determine thresholds for CV to classify items into X, Y, and Z categories. Common thresholds are as follows:
  - **X-items:** $CV \leq 0.1$ (low variability, predictable demand).
  - **Y-items:** $0.1 < CV \leq 0.25$ (moderate variability).
  - **Z-items:** $CV > 0.25$ (high variability, unpredictable demand).

- **Assign categories:** Based on the calculated CV, assign each item to the X, Y, or Z category.

Step 4: Analyze Results
- **X-items:** These items have stable and predictable demand. They require less safety stock and can be ordered in larger, less frequent batches.
- **Y-items:** These items show moderate variability. They may require seasonal or cyclical planning and moderate safety stock.
- **Z-items:** These items have highly variable and unpredictable demand. They need higher safety stock levels and more frequent monitoring.

3.4 XYZ analysis example

Let's say a retail store has three items with the following monthly demand data:

Item A:
- Monthly demand: [100, 102, 98, 101, 100, 99, 98, 102, 101, 100, 99, 100]
- Mean Demand = 100
- Standard Deviation = 1.42
- CV = 1.42 / 100 = 0.014 (X-item)

Item B:
- Monthly demand: [200, 190, 210, 195, 205, 190, 210, 205, 195, 200, 210, 190]
- Mean Demand = 200
- Standard Deviation = 7.94
- CV = 7.94 / 200 = 0.04 (X-item)

Item C:
- Monthly demand: [50, 100, 75, 150, 60, 110, 90, 70, 140, 55, 120, 80]
- Mean Demand = 90
- Standard Deviation = 33.17
- CV = 33.17 / 90 = 0.37 (Z-item)

3.5 Conducting ABC and XYZ analysis

Combining ABC and XYZ analyses provides a comprehensive approach to inventory management by considering both the value of items and the variability of their demand. This helps in creating more nuanced inventory strategies. Through it, businesses can develop a more refined inventory management strategy that considers both the financial importance and demand patterns of their inventory items. This holistic approach helps in optimizing stock levels, reducing costs, and improving overall efficiency. Steps that businesses can classify items using both ABC and XYZ analyses are as follows:

1. Conduct ABC Analysis — refer to previous section on **ABC Analysis Steps**.
2. Conduct XYZ Analysis — refer to previous section on **XYZ Analysis Steps**.
3. Combine both ABC and XYZ classifications.

Step 1: Create a Matrix — Create a 3 × 3 matrix with ABC categories on one axis and XYZ categories on the other axis.

	X	**Y**	**X**
A	A–X	A–Y	A–Z
B	B–X	B–Y	B–Z
C	C–X	C–Y	C–Z

Step 2: Place Items into Matrix — Assign each inventory item to a cell in the matrix based on its ABC and XYZ classifications.

	ABC Category	**XYZ Category**	**Combined Category**
High-End Smartphone	A	X	A–X
Mid-Range Laptop	A	X	A–Y
Smartwatch	B	Z	B–Z
Budget Earbuds	B	X	C–X
Phone Charger	C	Y	C–Y

4. Implement inventory strategies based on combined categories:
   - **A–X (high value, low variability):** Maintain consistent stock levels with regular replenishment cycles. Focus on efficiency and cost reduction.

- **A–Y (high value, moderate variability):** Monitor seasonal trends and adjust stock levels accordingly. Maintain a moderate level of safety stock.
- **A–Z (high value, high variability):** Implement high safety stock levels and frequent demand reviews. Prioritize reliable supply sources.
- **B–X (moderate value, low variability):** Standard replenishment practices with occasional reviews. Keep safety stock low.
- **B–Y (moderate value, moderate variability):** Seasonal adjustments and moderate safety stock levels. Regularly review demand trends.
- **B–Z (moderate value, high variability):** Higher safety stock levels and frequent monitoring. Responsive inventory adjustments.
- **C–X (low value, low variability):** Minimal safety stock. Order on a need basis to reduce holding costs.
- **C–Y (low value, moderate variability):** Adjust stock levels based on demand patterns with moderate safety stock.
- **C–Z (low value, high variability):** Minimal stock levels and frequent reviews. Consider just-in-time ordering.

3.6 *XYZ analysis advantages*

XYZ analysis is a powerful tool that helps businesses understand demand variability and optimize their inventory management practices. XYZ analysis offers several advantages that can significantly improve inventory management and demand forecasting. By leveraging these benefits, companies can achieve greater efficiency and customer satisfaction in their inventory operations.

1. **Improved demand forecasting:** XYZ analysis helps in categorizing items based on their demand variability, allowing businesses to apply different forecasting methods for each category. For example:
   - X-items (low variability) can use simple forecasting methods such as moving averages.
   - Y-items (moderate variability) may require seasonal adjustment models.
   - Z-items (high variability) might need advanced statistical methods or machine learning algorithms to predict demand accurately.

2. **Enhanced inventory control:** By understanding the demand patterns, companies can maintain optimal inventory levels:
   - X-items can maintain consistent stock levels with low safety stock.
   - Y-items can adjust inventory for seasonal trends with moderate safety stock.
   - Z-items can keep higher safety stock and closely monitor for sudden changes in demand.
3. **Cost reduction:** Accurate demand classification helps in reducing holding and shortage costs. Businesses can avoid overstocking items with predictable demand and understocking items with high demand variability by
   - minimizing excess inventory for X-items,
   - balancing inventory levels for Y-items,
   - increasing safety stock for Z-items to prevent stockouts.
4. **Better resource allocation:** XYZ analysis allows for more strategic resource allocation. High-priority resources and attention can be focused on items with high variability and value (e.g. A–Z items). This targeted approach ensures that critical items are managed more effectively.
5. **Improved service levels:** By aligning inventory strategies with demand variability, businesses can ensure better availability of products, leading to higher customer satisfaction and improved service levels. Customers are less likely to face stockouts, especially for high-variability items.
6. **Enhanced supplier management:** Understanding demand patterns helps in negotiating better terms with suppliers. For example:
   - Regular orders for X-items can lead to better pricing and consistent supply.
   - Flexible contracts for Y-items can accommodate seasonal variations.
   - More agile supplier arrangements for Z-items can ensure quick replenishment when demand spikes.
7. **Strategic decision-making:** XYZ analysis provides valuable insights that aid in strategic decision-making. For instance, businesses can decide
   - which items need more frequent reviews and adjustments,
   - where to invest in advanced demand forecasting tools,
   - how to optimize warehousing space based on demand predictability.

3.7 *XYZ analysis disadvantages*

While XYZ analysis is a valuable tool for understanding demand variability and optimizing inventory management, its disadvantages include the complexity of data collection and analysis, susceptibility to demand fluctuations, static nature of classification, neglect of other critical factors, and the resource-intensive nature of implementation. Businesses should carefully consider these drawbacks and potentially integrate XYZ analysis with other inventory management strategies to achieve more comprehensive results:

1. **Complexity in data collection and analysis:** XYZ analysis requires detailed historical demand data for accurate classification. Collecting, processing, and analyzing this data can be complex and time-consuming, especially for businesses with a large number of SKUs (Stock Keeping Units). The process involves calculating the mean, standard deviation, and coefficient of variation for each item, which demands robust data management systems and analytical tools.
2. **Demand fluctuations and market changes:** The analysis relies heavily on historical demand data, which may not always predict future demand accurately. Sudden market changes, shifts in consumer behavior, or unexpected external factors (such as economic downturns or global events) can make historical data less reliable. This can lead to inaccurate classifications and suboptimal inventory decisions.
3. **Static nature of classification:** XYZ analysis provides a static classification based on past data, which may not adapt quickly to changes in demand patterns. As a result, items may be misclassified over time if their demand variability changes. Regular updates and reclassifications are necessary to maintain accuracy, adding to the maintenance burden.
4. **Ignores other critical factors:** While XYZ analysis focuses on demand variability, it does not account for other important factors such as lead times, supplier reliability, or market trends. This narrow focus can result in incomplete inventory strategies that do not fully address all aspects of inventory management. Solely relying on demand variability might lead to suboptimal stocking decisions.
5. **Resource intensive:** Implementing XYZ analysis can be resource-intensive, requiring significant investment in data collection systems, analytical tools, and skilled personnel. Small businesses or those with limited resources may find it challenging to justify the cost and effort

involved. The need for continuous monitoring and updating of classifications further adds to the resource burden.

Self Check 2

1. How does XYZ analysis help businesses in optimizing their inventory management practices?
2. What are the key factors considered in classifying inventory items into X, Y, and Z categories in XYZ analysis?

---

4. SKU Rationalization and Standardization

SKU rationalization and standardization are inventory management strategies aimed at optimizing product offerings and improving operational efficiency. Both SKU rationalization and standardization aim to streamline inventory management processes, reduce costs, and improve overall efficiency. Rationalization focuses on optimizing the product mix, while standardization ensures consistency and accuracy in inventory data.

4.1 *SKU rationalization*

SKU rationalization is the process of evaluating and reducing the number of Stock Keeping Units (SKUs) in a company's inventory. The goal is to eliminate underperforming or redundant products that do not contribute significantly to sales or profitability. By streamlining the product assortment, businesses can reduce inventory carrying costs, improve cash flow, and focus on high-performing items. SKU rationalization involves analyzing sales data, customer preferences, and profitability metrics to identify which SKUs to retain or discontinue. This process helps reduce complexity in inventory management, improving forecasting accuracy, and enhancing overall supply chain efficiency.

4.2 *SKU standardization*

SKU standardization refers to the practice of ensuring that products are consistently identified and categorized using standardized SKU codes across the entire supply chain. This involves establishing uniform naming conventions, coding systems, and product attributes. Standardization simplifies inventory tracking, reduces errors, and enhances communication between different departments and stakeholders. It also facilitates better data analysis and reporting, leading to more informed decision-making. By adopting SKU standardization, companies can achieve greater consistency in their inventory records, improve operational efficiency, and enhance customer satisfaction through accurate and reliable product information.

4.3 *The need for SKU rationalization*

SKU rationalization is essential for businesses to maintain a streamlined, efficient, and profitable inventory management system. It is a critical strategy for optimizing inventory management. It helps reduce costs, improve cash flow, enhance operational efficiency, and increase customer satisfaction. By focusing on high-performing products and eliminating underper forming ones, businesses can better align their product offerings with market demand and achieve greater overall efficiency and profitability. Several key reasons why SKU rationalization is needed are as follows:

1. **Cost reduction:** Maintaining a large inventory of SKUs can be costly. These costs include not only the purchase price of the goods but also storage, handling, insurance, and potential obsolescence. By reducing the number of underperforming or redundant SKUs, businesses can significantly lower these carrying costs and improve their bottom line.
2. **Improved cash flow:** Holding excess inventory ties up capital that could be used more effectively elsewhere in the business. SKU rationalization helps free up this capital, allowing for investment in higher-performing products, new product development, or other strategic initiatives that can drive growth.
3. **Enhanced inventory management:** A streamlined SKU assortment simplifies inventory management tasks, such as ordering, stocking, and tracking. This leads to better accuracy in inventory counts, reduced stockouts or overstock situations, and more efficient use of warehouse space.
4. **Increased operational efficiency:** With fewer SKUs to manage, businesses can improve their operational efficiency. Simplified

processes reduce the complexity of order fulfillment, manufacturing, and distribution, leading to faster turnaround times and improved service levels.

5. **Better forecasting and demand planning:** A more focused product range makes it easier to predict demand and plan inventory levels accurately. This results in more reliable demand forecasts, reduced excess inventory, and minimized stockouts, ensuring that the right products are available when needed.

6. **Improved customer satisfaction:** SKU rationalization can enhance the customer experience by ensuring that the most popular and profitable products are always in stock. This leads to higher customer satisfaction and loyalty, as customers can rely on the availability of the products they want.

7. **Focus on core products:** By eliminating low-performing SKUs, businesses can concentrate their resources on developing and marketing their core product lines. This focus can lead to better product quality, more effective marketing strategies, and stronger brand identity.

8. **Alignment with market trends:** SKU rationalization allows businesses to stay agile and responsive to changing market conditions. By regularly reviewing and adjusting their product assortment, companies can quickly adapt to new trends, customer preferences, and competitive pressures.

4.4 *SKU rationalization types*

SKU rationalization can be approached through various strategies, each tailored to address specific business needs and goals. Different types of SKU rationalization serve various business objectives, from improving profitability and reducing redundancy to aligning with strategic goals and enhancing customer satisfaction. By adopting one or a combination of these rationalization strategies, businesses can streamline their product offerings, reduce costs, and better meet market demands. The main types of SKU rationalization are as follows:

1. **Performance-based rationalization:** This approach focuses on evaluating each SKU's performance based on sales volume, profitability, and contribution to overall revenue. SKUs that consistently underperform or fail to meet predefined performance criteria are considered for elimination. This type of rationalization helps businesses

concentrate on high-margin, high-volume products that drive profitability.

2. **Redundancy elimination:** Redundancy elimination involves identifying and removing duplicate or similar SKUs that serve the same purpose or target the same market segment. This can reduce complexity in inventory management, simplify customer choices, and lower carrying costs. For example, a company might consolidate multiple similar products into a single SKU with broader appeal.

3. **Seasonal and lifecycle rationalization:** Products with seasonal demand or those at different stages of their lifecycle (introduction, growth, maturity, and decline) require specific rationalization strategies. Seasonal rationalization involves adjusting the SKU assortment based on seasonal trends, ensuring that only relevant products are stocked during particular times of the year. Lifecycle rationalization focuses on phasing out SKUs that are in the decline stage and introducing new ones in the growth or introduction stages.

4. **Demand variability rationalization:** This method involves classifying SKUs based on their demand variability and predictability. SKUs with highly variable or unpredictable demand may be candidates for discontinuation or consolidation. This type of rationalization aligns inventory levels with actual demand patterns, reducing the risk of overstocking or stockouts.

5. **Strategic alignment rationalization:** Strategic alignment rationalization ensures that the SKU assortment aligns with the company's overall strategic goals and market positioning. This involves evaluating each SKU's relevance to the business's long-term strategy, brand identity, and target market. SKUs that do not align with strategic objectives may be phased out, even if they perform well individually.

6. **Profitability-based rationalization:** Profitability-based rationalization focuses on the financial performance of each SKU, considering factors such as gross margin, cost to serve, and total lifecycle cost. SKUs that are not profitable or have low margins may be discontinued, allowing the business to focus on more lucrative products.

7. **Customer-centric rationalization:** This approach centers around customer preferences and demand. By analyzing customer purchase patterns, feedback, and preferences, businesses can identify which SKUs are most valued by their customers and which are not. Rationalizing SKUs based on customer-centric data helps in retaining customer loyalty and satisfaction.

4.5 *Advantages and disadvantages of SKU rationalization*

SKU rationalization offers several advantages, including cost reduction, improved cash flow, enhanced inventory management, operational efficiency, better forecasting, increased customer satisfaction, focus on core competencies, and alignment with market trends. However, it also has disadvantages, such as potential loss of sales, impact on customer choice, implementation challenges, risk of incorrect decisions, and short-term disruptions. Businesses should carefully weigh these pros and cons when considering SKU rationalization to ensure they achieve the desired benefits without compromising customer satisfaction or operational stability.

Pros of SKU Rationalization	Cons of SKU Rationalization
Helps companies become more efficient and boosts productivity	Emphasizes efficiency at the expense of human capital
Allows management to implement modernized techniques and systems	Often involves large layoffs
Lowers market volatility	Can lead to a significantly increased workload for the workers who remain
Can provide the workforce with better working conditions and higher pay	Loss of initiative from workers due to the mechanization of processes
Translates into a higher standard of living in society	Costly and requires consistent monitoring
Can lead to lower prices and better products for consumers	No guarantee of improved returns

Self Check 3

1. How can SKU rationalization help a business improve its overall operational efficiency and reduce costs?
2. What are the key steps involved in implementing SKU standardization, and how does it contribute to better inventory management?

5. Summary

Chapter 3 delves into essential strategies for effective inventory management. ABC analysis is a method of categorizing inventory based on its importance, typically measured by the annual consumption value. Items are classified into three categories: A (high value), B (moderate value), and C (low value). This classification helps businesses prioritize their resources and management efforts on the most impactful items, ensuring optimal stock levels and efficient use of working capital. XYZ analysis focuses on classifying inventory based on demand variability and predictability. Items are divided into X (consistent demand), Y (variable demand), and Z (highly variable demand) categories. This analysis aids in tailoring demand forecasting and inventory strategies to the nature of demand fluctuations, thereby improving accuracy in demand planning and reducing the risks of overstocking or stockouts. SKU rationalization involves evaluating and reducing the number of Stock Keeping Units (SKUs) to eliminate underperforming or redundant products. This streamlining process reduces costs, improves cash flow, and enhances operational efficiency. SKU standardization, on the other hand, ensures consistent identification and categorization of products through standardized codes and naming conventions, facilitating better inventory tracking and data management. Together, these inventory classification and segmentation strategies enable businesses to optimize their inventory management, reduce costs, and align their product offerings with market demands and organizational goals.

6. Case Study: Effective Inventory Management at POSITIVE Electronics

POSITIVE Electronics is a mid-sized company specializing in consumer electronics. With a wide range of products, including smartphones, tablets, and accessories, the company faced challenges in managing its extensive inventory. Inefficient inventory practices led to high carrying costs, frequent stockouts, and excess inventory, ultimately affecting profitability and customer satisfaction. POSITIVE Electronics struggled with a few issues including overwhelming SKU proliferation, leading to complex inventory management, inaccurate demand forecasting, causing stockouts and overstock situations, and high inventory carrying costs and obsolescence rates. To address these issues, POSITIVE Electronics implemented a comprehensive inventory management strategy focusing on ABC analysis, XYZ analysis, and SKU rationalization and standardization.

The outcome from the ABC analysis was that POSITIVE Electronics focused resources and management efforts on A items, optimizing stock levels and reducing holding costs. Meanwhile, through XYZ analysis, POSITIVE Electronics was able to improve demand forecasting accuracy by tailoring inventory strategies to demand patterns, reducing stockouts and overstock situations. Whereby, through SKU rationalization and standardization, POSITIVE Electronics was able to reduce inventory carrying costs, improve cash flow, and enhance operational efficiency. Standardization facilitated better inventory tracking and data management. Overall, POSITIVE Electronics was able to achieve 15% reduction in inventory carrying costs and 25% improvement in demand forecasting accuracy and streamline inventory management processes, leading to faster order fulfillment and reduced complexity and increased product availability and reduced stockouts, enhancing customer satisfaction and loyalty. Hence, by implementing ABC analysis, XYZ analysis, and SKU rationalization and standardization, POSITIVE Electronics successfully optimized its inventory management. These strategies helped the company reduce costs, improve forecasting accuracy, and enhance operational efficiency, ultimately boosting profitability and customer satisfaction.

Case Study Questions

1. How did POSITIVE Electronics utilize ABC analysis to optimize its inventory management and reduce carrying costs?
2. In what ways did the application of XYZ analysis improve demand forecasting accuracy at POSITIVE Electronics?
3. What were the key outcomes of SKU rationalization and standardization for POSITIVE Electronics, and how did these strategies enhance operational efficiency?

References

Edward, A.S., David F.P., and Douglas J.T. (2021). *Inventory and Production Management in Supply Chains,* 4th edn. Boca Raton, Florida, USA: CRC Press.

Steven M.B. (2021). *Inventory Management,* 4th edn. Columbia, MD, USA: AccountingTools, Inc.

Vandeput, N. (2020). *Inventory Optimization: Models and Simulations*, 1st edn. Berlin, Germany: De Gruyter Publisher.

Chapter 4

Inventory Replenishment Strategies

Learning Outcome

By the end of this topic, you should be able to do the following:

1. Describe the just-in-time (JIT) inventory management.
2. Interpret the economic order quantity (EOQ) models.
3. Apply vendor-managed inventory (VMI).

1. Introduction

Chapter 4 delves into the pivotal for maintaining operational efficiency and cost-effectiveness in supply chain management. At its core, this chapter explores three fundamental approaches: Just-in-Time (JIT) inventory management, Economic Order Quantity (EOQ) models, and Vendor-Managed Inventory (VMI). First, the concept of JIT inventory management is explored, highlighting its essence in minimizing inventory holding costs while ensuring timely availability of goods. Understanding JIT entails understanding its principles of lean production, where inventory is replenished precisely when needed, thus streamlining processes and reducing waste. Next, this chapter explores the intricacies of EOQ models, providing a framework for determining optimal order quantities that strike a balance between inventory holding costs and ordering costs. Through EOQ analysis, businesses can fine-tune their replenishment strategies to achieve cost-efficiency without risking stockouts or excess inventory. Finally, VMI is explored as a collaborative approach between suppliers and buyers, where suppliers assume responsibility for managing inventory levels at customer locations. This symbiotic relationship optimizes inventory levels, enhances supply chain visibility, and fosters stronger partnerships between stakeholders. Learners will not only comprehend these inventory replenishment strategies but also establish their applicability and benefits within diverse operational contexts.

2. Just-In-Time (JIT) Inventory Management

2.1 *JIT inventory management overview*

JIT inventory management is a strategy that aligns raw-material orders from suppliers directly with production schedules. Its primary objective is to reduce inventory holding costs by receiving goods only as they are needed in the production process. This approach minimizes the amount of inventory on hand, thus lowering storage costs and reducing waste. For example, consider an automotive manufacturer that implements JIT. Instead of holding large inventories of car parts, the manufacturer receives parts such as tyres, engines, and seats from suppliers just as they are required for assembly. This not only reduces storage costs but also minimizes the risk of parts becoming obsolete. Another example can be seen in the fast-food industry. A popular fast-food chain such as McDonald's employs JIT principles by preparing food items only when an order is placed. Ingredients such as vegetables, buns, and patties are delivered to outlets frequently in small quantities to match the anticipated daily demand. This ensures freshness and reduces the space needed for storage. JIT inventory management requires robust coordination with suppliers and a reliable transportation network. It fosters a responsive and flexible production environment, though it also means that any disruption in the supply chain can halt production. Nevertheless, when executed effectively, JIT leads to significant cost savings and enhanced efficiency.

2.2 *JIT inventory management functions*

JIT inventory management relies on several core functions to ensure its successful implementation. These functions are crucial for minimizing waste, reducing costs, and maintaining a streamlined production process. The key function of JIT includes the following:

1. **Demand forecasting and planning:** Accurate demand forecasting is essential in JIT to align production schedules with customer demand. This involves using historical data, market trends, and predictive analytics to anticipate future demand and plan inventory accordingly.
2. **Supplier coordination:** Effective communication and coordination with suppliers are critical in JIT. Suppliers must be reliable and capable of delivering small quantities of materials at frequent intervals. Establishing strong relationships and maintaining transparent communication channels ensure timely deliveries.
3. **Inventory management:** JIT focuses on maintaining minimal inventory levels. This involves real-time tracking of inventory, regular audits, and efficient reordering processes. Inventory is ordered only when needed, reducing holding costs and minimizing the risk of obsolescence.
4. **Production scheduling:** Production scheduling in JIT is tightly synchronized with demand forecasts and inventory levels. This requires flexible production systems capable of quickly adapting to changes in demand without significant delays.
5. **Quality control:** Since JIT reduces buffer stock, quality control becomes even more critical. Ensuring high-quality standards at every stage of production helps prevent defects and reduces the need for rework, which can disrupt the JIT flow.
6. **Transportation and logistics:** Efficient transportation and logistics are vital for JIT. This includes optimizing delivery routes, ensuring timely shipments, and maintaining a reliable transportation network to prevent delays in the supply chain.
7. **Continuous improvement:** JIT promotes a culture of continuous improvement (Kaizen). Regularly reviewing processes, identifying inefficiencies, and implementing improvements help maintain the effectiveness of the JIT system.
8. **Employee training and involvement:** Employees play a crucial role in the success of JIT. Providing adequate training and involving employees in decision-making processes enhance their ability to identify issues, suggest improvements, and maintain high productivity levels.

2.3 *Eight types of wastes*

In the context of JIT inventory management and lean manufacturing, identifying and eliminating waste is crucial to enhancing efficiency and productivity. By identifying and addressing these eight types of wastes, organizations can streamline operations, reduce costs, and improve over-all efficiency, ultimately delivering greater value to customers. There are eight types of wastes commonly referred to as "Muda" in lean terminology. The eight types of wastes are listed in Table 1:

Table 1. Type of wastes under JIT.

Waste	Definition	Example
Overproduction	Producing more than what is needed or before it is needed.	A bakery producing more bread than the daily sales demand, resulting in excess inventory that may go stale.
Waiting	Idle time when resources are not being used efficiently.	Assembly line workers waiting for materials to arrive because of a delay from the supplier.
Transport	Unnecessary movement of products or materials.	Frequently moving parts from one side of the warehouse to another without any added value to the product.
Extra processing	Performing more work or adding more features than what is required by the customer.	Polishing a part of a machine that will not be visible or contribute to its functionality.
Inventory	Holding more inventory than necessary leads to increased holding costs and potential obsolescence.	A clothing retailer stocking excessive amounts of seasonal clothing, risking that unsold items will go out of fashion.
Motion	Unnecessary movement by people within the workspace.	Workers walk long distances between their workstations and the storage area to retrieve tools or parts.
Defects	Producing defective products that require rework or scrapping.	A manufacturing defect in a smartphone leads to a recall and the need for repairs, costing time and money.
Non-utilized talent	Underutilizing employees' skills and capabilities.	Highly skilled workers are assigned to menial tasks that do not fully utilize their expertise, leading to disengagement and lost potential.

2.4 *Strategies for minimizing wastes by using JIT*

Implementing JIT inventory management is an effective way to minimize waste in manufacturing and other operational processes. By integrating these JIT strategies, organizations can effectively minimize waste, streamline processes, and enhance overall operational efficiency.

Some of the strategies for minimizing the eight types of wastes using JIT principles are listed in Table 2:

Table 2.   Strategies to minimize wastes.

Waste	Strategy	Example
Overproduction	Produce based on actual customer demand rather than forecasts.	Use JIT to align production schedules with real-time sales orders, ensuring products are manufactured only when needed. Implement a pull system where production is triggered by customer demand.
Waiting	Streamline workflows and enhance coordination with suppliers.	Implement JIT scheduling to synchronize supply deliveries with production cycles, reducing downtime. Use Kanban systems to signal when materials are needed, ensuring continuous production flow.
Transport	Optimize the layout of production facilities and streamline logistics.	Design the production floor to minimize the distance materials and products travel. Use JIT to coordinate direct delivery of materials to the point of use, reducing unnecessary handling and transportation.
Extra Processing	Focus on value-added activities and eliminate unnecessary steps.	Conduct process analysis to identify and remove redundant processes. Use JIT to ensure that every production step adds value and meets customer specifications without overprocessing.
Inventory	Maintain minimal inventory levels through precise demand forecasting and supplier collaboration.	Implement JIT inventory practices to order materials in smaller, more frequent batches based on actual production needs. Use supplier partnerships to ensure timely deliveries, reducing the need for excess inventory.

(Continued)

Table 2.   (*Continued*)

Waste	Strategy	Example
Motion	Optimize workspace organization and ergonomics.	Implement JIT principles to design efficient workstations where tools and materials are within easy reach. Use 5S (Sort, Set in order, Shine, Standardize, Sustain) methodology to create organized and efficient work environments.
Defects	Enhance quality control and focus on defect prevention.	Use JIT to implement real-time quality checks and immediate feedback loops in the production process. Employ continuous improvement (Kaizen) practices to identify and address root causes of defects, reducing rework and scrap.
Non-Utilized Talent	Engage and empower employees by leveraging their skills and insights.	Use JIT to create cross-functional teams and involve employees in problem-solving and process improvement initiatives. Provide training and development opportunities to fully utilize and enhance employees' capabilities.

Implementation example: A car manufacturer using JIT receives parts from suppliers exactly when needed on the assembly line, reducing inventory costs. Production is closely tied to customer orders, avoiding overproduction. Workers are cross-trained to perform multiple tasks, minimizing waiting time and maximizing their skills. Regular feedback and quality checks catch defects early, preventing rework. The factory layout is optimized to reduce unnecessary movement, ensuring smooth and efficient operations.

2.5 *JIT practices: A case study*

Toyota, Dell, and Harley-Davidson exemplify successful JIT implementation, each adapting the strategy to their unique operational needs. Toyota's rigorous supplier integration and Kanban system, Dell's build-to-order model and real-time supplier collaboration, and Harley-Davidson's continuous improvement and flexible production systems demonstrate the versatility and benefits of JIT. Common benefits across these companies include reduced inventory costs, improved quality control, and enhanced operational efficiency, leading to competitive advantages and increased profitability as listed in Table 3.

Table 3.   JIT implementation by Toyota, Dell, and Harley-Davidson.

Company	Strategies	Benefits
Toyota	Toyota is the pioneer of JIT inventory management, implementing it as part of the Toyota Production System (TPS). Its JIT strategy focuses on producing only what is needed, when it is needed, and in the amount needed. Key elements include the following: 1. **Kanban system:** Visual signals are used to trigger the movement of materials within the production process and to signal the need for more inventory. 2. **Supplier integration:** Toyota maintains close relationships with suppliers, ensuring timely deliveries of components. 3. **Production smoothing (Heijunka):** Production schedules are leveled to avoid fluctuations, leading to a more consistent flow of work.	1. **Reduced inventory costs:** Holding costs are minimized as inventory levels are kept low. 2. **Enhanced quality control:** Defects are detected and corrected promptly, improving overall product quality. 3. **Increased efficiency:** Streamlined production processes reduce waste and improve operational efficiency.
Dell	Dell revolutionized the personal computer industry with its direct-to-consumer sales model, tightly coupled with JIT principles. Dell's approach includes the following: 1. **Build-to-order:** Computers are assembled only after a customer places an order, allowing for customization and minimizing excess inventory.	1. **Cost reduction:** Reduced inventory carrying costs and minimal obsolescence of components. 2. **Customization:** Ability to offer customized products tailored to individual customer needs.

(Continued)

Table 3.   (*Continued*)

Company	Strategies	Benefits
	2. **Supplier hubs:** Dell's suppliers maintain inventory close to Dell's assembly plants, enabling quick replenishment. 3. **Integrated supply chain:** Real-time data sharing with suppliers ensures components are delivered just in time for production.	3. **Speed to market:** Faster assembly and delivery times enhance customer satisfaction and competitiveness.
Harley-Davidson	Harley-Davidson implemented JIT to revitalize its manufacturing process and compete more effectively in the global market. Its JIT strategies include the following: 1. **Continuous improvement (Kaizen):** Ongoing efforts to improve production processes and eliminate waste. 2. **Employee involvement:** Empowering employees to take part in decision-making and problem-solving to enhance productivity. 3. **Flexible production systems:** Adopting modular manufacturing systems that allow for quick adjustments to production schedules based on demand.	1. **Improved production efficiency:** Reduction in lead times and increased flexibility in manufacturing. 2. **Enhanced product quality:** Better quality control and reduced defect rates. 3. **Cost Savings:** Lower inventory levels and reduced warehousing costs contribute to significant savings.

2.6 *JIT inventory management benefits*

JIT inventory management offers a range of benefits that enhance efficiency, reduce costs, and improve overall operational performance. By adopting JIT inventory management, companies can achieve a leaner, more responsive, and cost-effective production environment, ultimately leading to competitive advantages and sustainable growth. Some of the key benefits of implementing JIT are as follows:

1. **Reduced inventory costs:** JIT minimizes the amount of inventory held at any given time, significantly reducing storage and holding

costs. By ordering and receiving goods only as needed, companies avoid the expenses associated with maintaining large inventories, such as warehousing, insurance, and spoilage costs.

2. **Decreased waste:** By aligning production closely with demand, JIT helps reduce waste in several forms, including overproduction, excess inventory, and obsolescence. This lean approach ensures that resources are used efficiently and only in necessary quantities.

3. **Improved cash flow:** With less capital tied up in inventory, companies can better manage their cash flow. This allows for more flexibility in financial planning and investment in other areas of the business, such as research and development or marketing.

4. **Enhanced quality control:** JIT emphasizes continuous improvement and immediate feedback, leading to better quality control. Defects are identified and corrected promptly, reducing the incidence of faulty products and enhancing overall product quality.

5. **Increased efficiency:** Streamlined production processes and synchronized supply chains lead to increased operational efficiency. JIT reduces downtime by ensuring that materials and components are available precisely when needed, avoiding delays and bottlenecks in production.

6. **Greater flexibility:** JIT allows companies to respond quickly to changes in customer demand and market conditions. By maintaining minimal inventory and flexible production systems, businesses can adapt more easily to fluctuations without the burden of excess stock.

7. **Better supplier relationships:** Implementing JIT often involves closer collaboration with suppliers. This can lead to stronger relationships, better communication, and more reliable delivery schedules. Suppliers become partners in the process, working together to achieve mutual goals.

8. **Higher customer satisfaction:** With JIT, companies can deliver products more quickly and reliably. Reduced lead times and the ability to meet customer-specific demands improve overall customer satisfaction and loyalty.

9. **Space utilization:** By reducing the need for large storage spaces, JIT allows companies to utilize their facilities more efficiently. Space that was previously used for storing excess inventory can be repurposed for other productive activities.

10. **Continuous improvement:** JIT fosters a culture of continuous improvement (Kaizen) by encouraging regular evaluation and optimization of processes. This ongoing focus on improvement helps maintain high standards of efficiency and quality.

2.7 *JIT disadvantages*

JIT inventory management offers numerous benefits, but it also comes with several disadvantages. Companies must carefully manage and mitigate these disadvantages to ensure smooth and efficient operations. Balancing the benefits of reduced inventory with the risks of supply chain disruptions, increased costs, and demand fluctuations is essential for successful JIT implementation. The five main disadvantages of JIT are as follows:

1. **Vulnerability to supply chain disruptions:** JIT relies heavily on the timely delivery of materials and components. Any disruption in the supply chain, such as delays from suppliers, transportation issues, or natural disasters, can halt production. Since there is little to no buffer inventory, companies may face significant downtime and lost productivity if a supply chain disruption occurs.

2. **Increased supplier dependence:** Implementing JIT requires a high level of coordination and reliability from suppliers. Companies become more dependent on their suppliers for timely deliveries. If a supplier fails to meet delivery schedules, it can lead to production delays and potentially damage customer relationships. This increased dependency can also reduce a company's bargaining power with suppliers.

3. **Higher transaction costs:** Frequent ordering of smaller quantities, which is a core aspect of JIT, can lead to higher transaction costs. Administrative tasks, such as processing orders and handling invoices, are required more frequently, which can increase operational costs. Additionally, frequent transportation can raise logistics costs, particularly if suppliers are located far from the production facility.

4. **Limited economies of scale:** JIT reduces inventory levels and often involves ordering smaller quantities more frequently, which can prevent companies from taking advantage of bulk purchasing discounts. This lack of economies of scale can result in higher per-unit costs for materials and components, potentially increasing overall production costs.

5. **Inflexibility in responding to demand surges:** While JIT is designed to match production closely with demand, it can be challenging to quickly scale up production in response to sudden demand surges. With minimal inventory on hand and a production system tuned to

meet current demand, companies may struggle to fulfill large, unexpected orders promptly. This inflexibility can lead to lost sales and decreased customer satisfaction.

Self Check 1

1. What are the primary benefits of implementing Just-in-Time (JIT) inventory management, and how do these benefits enhance operational efficiency and customer satisfaction in a manufacturing environment?
2. Discuss the main disadvantages of Just-in-Time (JIT) inventory management. How can companies mitigate the risks associated with supply chain disruptions and increased supplier dependence when using JIT?

---

3. Economic Order Quantity (EOQ) Models

3.1 *What is EOQ?*

EOQ is a fundamental concept in inventory management that determines the optimal order quantity a company should purchase to minimize the total costs associated with inventory management. These costs include both ordering costs (the costs of placing and receiving orders) and holding costs (the costs of storing and maintaining inventory). EOQ is a valuable metric for businesses involved in purchasing and holding inventory for manufacturing, resale, internal use, or other purposes. Companies utilizing EOQ consider all costs associated with purchasing and delivery while also accounting for product demand, purchase discounts, and holding costs.

Savvy business owners and managers recognize the complexity of determining ideal inventory levels. When vendors provide volume discounts and other purchasing incentives, EOQ can assist in finding the optimal balance. EOQ is a powerful tool for businesses to manage their inventory effectively. By calculating the ideal order quantity, companies can achieve significant cost savings, improve inventory turnover, and enhance overall operational efficiency. EOQ uses the economic order quantity formula (provided in the following) to deliver a data-driven result that helps optimize business profitability. The EOQ is calculated using the following formula:

$$EOQ = \sqrt{\frac{2DS}{H}}$$

where D is the annual demand for the product, S is the ordering cost per order, and H is the holding cost per unit per year.

3.2 *EOQ assumption*

EOQ relies on several key assumptions to provide accurate and useful results. These assumptions simplify the real-world complexities of inventory management, making the EOQ model more practical for theoretical and analytical purposes. The primary assumptions underlying the EOQ model are as follows:

1. **Constant demand:** The EOQ model's assumption of constant demand implies that the product's sales rate remains steady and predictable over time, without any peaks or troughs. This simplifies inventory calculations, as it eliminates the need to account for variable demand patterns, such as seasonal spikes or sudden drops.
2. **Constant lead time:** The EOQ model assumes that lead time (the duration between placing an order and receiving it) remains consistent and predictable. This means there are no unexpected delays or variations in delivery times, allowing businesses to plan inventory replenishment accurately without the risk of stockouts or excess inventory due to timing uncertainties.
3. **Constant ordering cost:** The EOQ model assumes that the ordering cost is constant, meaning it remains the same regardless of the order size. This fixed cost includes all expenses related to placing an order, such as administrative tasks, transportation, and handling fees, simplifying the calculation of total ordering expenses.
4. **Constant holding cost:** The EOQ model assumes that the holding cost per unit is constant and predictable throughout the year. This cost encompasses expenses such as storage, insurance, and the opportunity cost of capital tied up in inventory, allowing for straightforward calculations in determining the optimal order quantity.

5. **Instantaneous replenishment:** The EOQ model assumes instantaneous replenishment, meaning that inventory is immediately restocked upon placing an order. This eliminates the risk of stockouts or backorders, as there are no delays between ordering and receiving inventory, ensuring a continuous supply of products without interruptions.

6. **No quantity discounts:** The EOQ model assumes a constant unit price, ignoring any potential savings from bulk purchasing or volume discounts. This simplification means that the cost per item remains unchanged regardless of the order size, allowing the model to focus solely on balancing ordering and holding costs without considering price variations.

7. **Single product:** The EOQ model is designed for a single product, ignoring complexities arising from managing multiple items. It does not consider interactions such as shared storage space or correlated demand between different products, simplifying inventory calculations by focusing on optimizing order quantities for just one product at a time.

8. **Full visibility and accuracy:** The EOQ model presumes complete and accurate data on-demand rates, ordering costs, and holding costs. It assumes no uncertainties or errors in this information, allowing for precise calculations. Any inaccuracies or lack of data could lead to suboptimal inventory decisions and affect overall effectiveness.

3.3 *Example scenario*

Consider a small electronics retailer using EOQ to manage its inventory of smartphone accessories. The retailer assumes that it sells 1,000 units of a specific accessory each year (constant demand), that it takes exactly one week to receive an order from the supplier (constant lead time), and that the cost to place each order is $50 (constant ordering cost). The retailer also knows that it costs $1 per unit per year to hold inventory (constant holding cost). The EOQ model can then be used to determine the optimal order quantity to minimize total inventory costs, under these simplifying assumptions.

While the EOQ model provides a straightforward and useful tool for optimizing inventory management, it is important to recognize its

assumptions. Real-world scenarios may require adjustments and considerations beyond these basic assumptions to accurately reflect the complexities of actual business operations.

3.4 *EOQ examples*

Consider a bookstore that sells 12,000 copies of a particular book annually. The cost to place an order (ordering cost) is $30, and the cost to hold
one book in inventory for a year (holding cost) is $2. Using the EOQ
formula,

$$EOQ = \sqrt{\frac{2 \times 12,000 \times 30}{2}}$$
$$EOQ = \sqrt{36,000}$$
$$EOQ = 600$$

Hence, the bookstore should order 600 copies of the book each time
they place an order to minimize their total inventory costs.

3.5 *EOQ purpose*

The primary goal of EOQ is to identify the most economical quantity of
inventory to order, balancing the costs associated with ordering and holding inventory. By achieving this balance, businesses can reduce their total
inventory costs and improve operational efficiency. These costs include
the following:

1. **Ordering costs:** Expenses related to placing and receiving orders,
   such as administrative fees and transportation costs.
2. **Holding costs:** Costs incurred from storing inventory, including
   warehousing, insurance, and opportunity costs of capital tied up in
   stock.

By calculating the EOQ, businesses aim to achieve a balance between
these costs, avoiding both excessive inventory (which increases holding
costs) and frequent reordering (which raises ordering costs). EOQ helps
businesses streamline their inventory processes, improve efficiency, and

enhance profitability. Specifically, the EOQ model helps in the following:

1. **Reducing total inventory costs:** Optimizing order quantities with EOQ helps businesses minimize total inventory costs by balancing ordering and holding expenses. By determining the most cost-effective order size, companies reduce excess stock and frequent orders, leading to lower overall costs associated with managing inventory and improving financial efficiency.
2. **Improving cash flow:** By maintaining lower inventory levels through EOQ, businesses reduce the amount of capital invested in stock. This freed-up capital can be redirected to other productive areas, such as investment in new projects or expansion opportunities, enhancing overall cash flow and financial flexibility.
3. **Enhancing inventory management:** EOQ offers a structured method for ordering inventory that matches supply with demand. This systematic approach minimizes the risk of stockouts by ensuring sufficient stock levels and reduces excess inventory by avoiding overordering, leading to more efficient inventory management and better alignment with actual consumption patterns.

3.6 *EOQ benefits*

The EOQ model provides several benefits to businesses by optimizing inventory management. By leveraging EOQ, businesses can achieve a more balanced, efficient, and cost-effective approach to inventory management, ultimately supporting better operational and financial outcomes. Some key advantages include the following:

1. **Cost reduction:** EOQ reduces total inventory costs by finding the ideal order size that minimizes both ordering and holding expenses. For example, a company that sells 5,000 units annually might incur $100 per order and $2 per unit in holding costs. Using EOQ, the company finds the optimal order quantity of 500 units. This minimizes the combined costs of frequent ordering and high storage fees, resulting in lower overall expenses compared to ordering in larger or smaller quantities.
2. **Improved inventory turnover:** EOQ improves inventory turnover by determining the optimal order quantity, which prevents

overstocking and understocking. For instance, a bookstore using EOQ to order 200 copies of a popular novel avoids having too many unsold books (which could become obsolete) or too few (leading to missed sales). By aligning inventory levels with actual demand, the bookstore maintains a steady flow of sales, reduces excess inventory, and ensures efficient stock replenishment, enhancing overall inventory management.

3. **Better cash flow management:** EOQ enhances cash flow by reducing inventory levels, which means less capital is locked up in stock. For example, a manufacturing company that uses EOQ to order 1,000 units instead of 2,000 frees up $20,000 previously tied in excess inventory. This released capital can be reinvested in expanding production or upgrading technology. As a result, the company improves its financial flexibility and operational efficiency, enabling growth and development without straining its cash resources.

4. **Increased operational efficiency:** EOQ increases operational efficiency by standardizing the order process, making inventory management more predictable. For example, a retailer using EOQ to order 300 units per order instead of varying quantities each time streamlines inventory planning. This consistency reduces the need for frequent adjustments and minimizes administrative tasks related to order placement and inventory tracking. As a result, the retailer experiences smoother operations, lower administrative costs, and fewer logistical issues, leading to overall improved efficiency in managing stock levels.

5. **Enhanced decision-making:** EOQ enhances decision-making by providing a data-driven method for inventory management. For example, an electronics retailer calculates EOQ and determines that ordering 400 units each time is optimal. This data-driven insight allows the retailer to plan inventory purchases more accurately, avoiding both surplus and shortages. With precise data on order size and inventory costs, the retailer can make informed decisions about stock levels, budgeting, and supplier negotiations, leading to more effective and strategic inventory control.

6. **Reduced stockouts and overstocks:** EOQ helps balance inventory levels, reducing the risks of stockouts and overstocks. For example, a toy store using EOQ to order 150 units of a popular toy ensures it has enough stock to meet demand during the holiday season without overstocking. By avoiding stockouts, the store prevents missed sales, and by preventing overstocks, it reduces excess inventory and storage

costs. This optimal order quantity ensures that the toy is available when needed while minimizing surplus and related expenses.

7. **Improved supplier relationships:** EOQ fosters improved supplier relationships by establishing consistent and predictable ordering patterns. For example, a coffee shop using EOQ to order 100 bags of coffee beans monthly maintains a steady order schedule. This predictability enables the supplier to plan production and delivery more efficiently. As a result, the coffee shop can negotiate better terms, such as discounts or priority service, and the supplier benefits from a reliable customer, leading to a mutually beneficial, long-term partnership.

8. **Enhanced customer satisfaction:** EOQ enhances customer satisfaction by ensuring optimal inventory levels to meet demand. For instance, an online retailer using EOQ to order 500 units of a popular gadget ensures it has enough stock to fulfill customer orders promptly. By consistently having the product available, the retailer avoids stockouts and delays, providing reliable service. This availability boosts customer satisfaction and loyalty, leading to positive reviews and repeat business, which in turn strengthens the brand's reputation and customer trust.

3.7 *EOQ limitations and challenges*

EOQ is a valuable tool, but it has limitations and challenges that can impact its effectiveness in real-world scenarios. Addressing these limitations often requires integrating EOQ with other inventory management techniques and adjusting strategies based on specific business needs and market conditions. Some of the key limitations and challenges are as follows:

1. **Assumption of constant demand:** EOQ's assumption of constant demand simplifies inventory management by assuming a steady, predictable sales rate. However, in reality, demand often fluctuates due to seasonal trends, market shifts, or unforeseen events. For example, a retailer might experience higher demand during holidays and lower demand in off-peak seasons. EOQ does not adjust for these variations, which can lead to either stockouts during peak times or excess inventory during slow periods. To address this challenge, businesses may need to combine EOQ with dynamic inventory models or demand forecasting techniques.

2. **Fixed ordering and holding costs:** EOQ assumes that ordering and holding costs remain constant, which simplifies calculations. In practice, these costs can fluctuate due to changes in supplier pricing, storage fees, or logistical expenses. For instance, a rise in transportation costs or increased warehouse rental fees can impact the total cost of inventory management. If these variations are not accounted for, the EOQ model may suggest suboptimal order quantities, leading to increased costs or inefficiencies. Businesses must regularly review and adjust their EOQ calculations to reflect changes in these costs for accurate inventory management.

3. **No consideration for quantity discounts:** EOQ assumes a fixed unit price, not considering potential discounts for bulk purchases. For example, a supplier might offer lower prices for orders above 1,000 units. If a business follows EOQ and orders less than this threshold, it may miss out on cost savings from bulk discounts. This can lead to higher overall inventory costs compared to ordering larger quantities at discounted rates. To address this challenge, businesses should integrate discount opportunities into their inventory models, possibly using mixed strategies that balance EOQ with volume pricing benefits.

4. **Simplified inventory model:** EOQ is designed for single-product inventory management, simplifying calculations by focusing on one item at a time. However, businesses with diverse product lines or interdependent inventory needs face more complex challenges. For example, a retailer managing multiple products with varying demand patterns and storage requirements may find EOQ insufficient, as it doesn't account for interactions between products or shared resources. To manage such complexities, companies often need integrated inventory systems or multi-product models that consider the relationships and dependencies among different items.

5. **Instantaneous replenishment assumption:** EOQ assumes that inventory replenishment is instantaneous, which does not account for real-world lead times or potential supply chain disruptions. For example, if a supplier experiences delays or if the lead time extends beyond the expected period, a business may face stockouts or excessive inventory if relying solely on EOQ. These issues arise because EOQ calculations assume immediate stock availability, making them less reliable when actual lead times vary. To mitigate this, businesses should integrate EOQ with real-time inventory tracking and adjust reorder points based on actual lead times.

6. **Requires accurate data:** EOQ relies on precise data for demand rates, ordering costs, and holding costs to calculate optimal order quantities. If this data is inaccurate or incomplete, EOQ calculations can lead to incorrect order sizes. For instance, if demand rates are overestimated, a business might order too much, resulting in excess inventory and higher holding costs. Conversely, underestimating demand can lead to stockouts and missed sales. Reliable and updated data is crucial for EOQ to function effectively; otherwise, it can compromise inventory management and increase overall costs.

7. **Not suitable for all industries:** EOQ may not be effective for industries with highly perishable goods or rapid inventory turnover, such as fresh food or high-demand electronics. For example, a grocery store managing perishable items such as fruits requires frequent replenishment to prevent spoilage, making EOQ's fixed order quantity less practical. Similarly, businesses with fast-moving inventory need more dynamic models to adjust quickly to changing demand. In these cases, alternative methods such as just-in-time (JIT) or dynamic replenishment systems are often more suitable, catering to the unique challenges of managing perishable or fast-moving items.

8. **Complex supply chain dynamics:** EOQ does not consider complex supply chain dynamics such as supplier reliability or market volatility. For instance, if a supplier frequently delays shipments or if market conditions cause price fluctuations, the EOQ model's fixed order quantities may become ineffective. These variations can lead to stockouts or excess inventory, as EOQ assumes a stable supply chain. To address these challenges, businesses must incorporate supply chain risk management strategies, such as safety stock, diversified suppliers, or flexible order quantities, to handle uncertainties and maintain effective inventory control.

Self Check 2

1. How does the assumption of constant demand in the EOQ model affect its applicability to businesses with seasonal or fluctuating demand patterns?

2. What strategies can businesses use to address the limitations of EOQ in industries with rapid inventory turnover or highly perishable goods?

4. Vendor-Managed Inventory (VMI)

4.1 *VMI defined*

VMI is a collaborative inventory management approach where the vendor assumes responsibility for managing the retailer's inventory. In this system, the vendor monitors inventory levels and determines when to replenish stock, eliminating the retailer's need to place orders actively. The primary objective of VMI is to minimize inventory-related costs for both parties. By having the vendor manage inventory, the stock is replenished only as needed, reducing the risk of overstock and lowering costs for the retailer. For the vendor, VMI simplifies operations by creating a more predictable and efficient inventory flow.

4.2 *How does a VMI system work?*

In a VMI setup, effective communication between the seller and the buyer is essential. Initially, both parties establish success metrics and agree on partnership terms. The vendor then begins shipping products to the retailer, who provides ongoing inventory data to the vendor. This data allows the vendor to monitor stock levels and purchasing trends accurately, which is crucial for VMI's success. With the data provided, the vendor manages inventory replenishment, relieving the retailer from ordering pressures. The vendor's insight into production schedules, shipping delays, and lead times helps optimize inventory flow and cost savings. Payment for the inventory follows an agreement between the retailer and vendor, which may stipulate payment upon stock arrival or upon sale. For example, a pet store chain utilizing VMI for dog food inventory would have the dog food supplier manage stock levels. The supplier uses regular sales data from the pet store to determine when and how much dog food to restock, based on pre-established goals. VMI is particularly advantageous for retailers managing a diverse product range, as it simplifies inventory management by transferring responsibility to the vendor. VMI stands out from other inventory management systems due to the following three key factors:

1. **Shared information:** Effective inventory management relies on the exchange of crucial data between businesses and retailers. This shared information includes real-time inventory levels, sales trends, and

order history. Accurate data ensure the vendor can make informed decisions about stock replenishment, optimizing inventory levels and reducing stockouts or overstocking.

2. **Vendor control:** Vendors are responsible for managing inventory, including monitoring stock levels, forecasting demand, and scheduling replenishments. This control allows vendors to optimize inventory flow based on real-time data, reducing the retailer's burden and ensuring that stock is replenished efficiently without requiring the retailer's direct involvement.

3. **Vendor responsibilities:** Vendors manage all aspects of restocking and ordering. This includes determining when and how much inventory to reorder based on real-time data from the retailer. By handling these duties, vendors ensure timely replenishment and minimize the retailer's administrative workload, leading to more efficient inventory management.

4.3 *Benefits of VMI*

VMI offers several benefits for both vendors and retailers, including the following:

1. **Reduced inventory costs:** By taking control of inventory levels, vendors ensure optimal stock quantities, reducing the retailer's need to hold excess inventory. This minimizes holding costs and lowers the risk of overstocking, leading to cost savings. Efficient stock management by vendors helps avoid unnecessary inventory expenses for retailers.

2. **Improved stock availability:** Vendors use real-time data to manage inventory, ensuring stock levels are closely aligned with actual demand. This proactive approach minimizes the risk of stockouts, so products remain available when customers need them. Improved stock availability enhances customer satisfaction and helps retailers maintain consistent sales.

3. **Streamlined operations:** Vendors manage all aspects of inventory replenishment and ordering, freeing retailers from these tasks. This streamlined approach reduces the retailer's administrative workload, allowing them to focus on other business areas. By automating and optimizing these processes, overall operational efficiency and effectiveness are significantly improved.

4. **Enhanced collaboration:** VMI promotes closer collaboration between vendors and retailers by sharing inventory data and maintaining open communication. This partnership leads to better alignment on inventory needs, demand forecasting, and replenishment strategies, enhancing coordination and fostering a more productive and efficient business relationship.

5. **Better demand forecasting:** With access to real-time sales and inventory data, vendors can forecast demand with greater accuracy. This allows for precise inventory planning, minimizing the risk of overstocking and stockouts. Accurate demand forecasting ensures that inventory levels are aligned with actual market needs, improving overall inventory efficiency.

6. **Increased efficiency:** Automated replenishment in VMI streamlines inventory management, leading to faster and more accurate order fulfillment. This efficiency reduces lead times and enhances supply chain operations. Improved inventory control minimizes delays and optimizes stock flow, boosting overall performance and operational effectiveness throughout the supply chain.

7. **Cost savings:** VMI optimizes inventory levels, reducing the frequency of stockouts and excess stock. This efficiency leads to cost savings by lowering ordering costs and minimizing markdowns on unsold inventory. Both vendors and retailers benefit from reduced financial losses and improved profitability through better inventory management.

4.4 *Drawbacks of VMI*

VMI offers many benefits but also has its drawbacks. The five main drawbacks are as follows:

1. **Dependency on vendor:** Retailers depend on vendors to manage inventory. For instance, if a vendor miscalculates stock levels or delays replenishment, it can lead to stockouts for the retailer, affecting sales. For example, if a grocery store relies on a vendor for stocking fresh produce and the vendor fails to deliver on time, the store may run out of products, leading to lost sales and customer dissatisfaction.

This dependency highlights the risk of vendor-related issues impacting the retailer's operations and revenue.

2. **Data security concerns:** Sharing inventory and sales data with vendors in a VMI system raises data security concerns. For example, if a retailer's sales data is accessed by an unauthorized party or if the vendor's system is compromised, sensitive information could be leaked or misused. This could expose the retailer's strategic data, potentially leading to competitive disadvantages or financial loss. For instance, a vendor could unintentionally or maliciously share pricing strategies with competitors, harming the retailer's market position and profitability.

3. **Complexity in implementation:** Implementing VMI involves complex coordination between retailers and vendors. For example, integrating different IT systems for real-time data sharing can be challenging and time-consuming. Retailers and vendors must align their inventory management processes and negotiate detailed contracts outlining responsibilities and performance metrics. This complexity was evident when a large retailer struggled to synchronize its ERP system with a vendor's inventory system, causing delays in inventory data updates and affecting the efficiency of the replenishment process.

4. **Reduced control:** In VMI, retailers might have reduced control over inventory decisions. For example, if a fashion retailer relies on a vendor to manage stock levels, the vendor might prioritize replenishing popular items based on their own metrics, potentially neglecting niche or seasonal products important to the retailer. This misalignment can lead to stock imbalances, where the retailer's specific preferences or local market needs are not fully addressed, impacting product availability and customer satisfaction.

5. **Potential for over-reliance:** Over-reliance on a vendor in a VMI system can create vulnerabilities. For instance, if a vendor experiences supply chain disruptions or financial difficulties, it could result in delayed deliveries or inventory shortages for the retailer. For example, a clothing retailer relying on a single vendor for seasonal merchandise might face empty shelves if the vendor encounters production issues. This dependency highlights the risk of operational problems at the vendor level directly impacting the retailer's inventory management and sales.

4.5 *VMI best practices*

Implementing VMI effectively requires adherence to several best practices. The main best practices for implementing VMI that the company should consider are as follows:

1. **Establish clear goals and metrics:** Define specific objectives for the VMI partnership, such as reducing stockouts or improving order accuracy. Set measurable performance metrics to track progress and ensure both parties are aligned with these goals.
2. **Ensure accurate data sharing:** Provide real-time and accurate inventory and sales data to the vendor. Reliable data is crucial for the vendor to forecast demand and manage inventory effectively. Regularly validate and update this data to maintain its accuracy.
3. **Develop strong communication:** Maintain open and consistent communication channels between the retailer and vendor. Regular updates, meetings, and feedback sessions help address issues promptly, align strategies, and enhance collaboration.
4. **Align processes and systems:** Integrate inventory management systems between the retailer and vendor to facilitate seamless data exchange and automated replenishment. Ensure that both systems are compatible to streamline operations and reduce manual errors.
5. **Negotiate clear agreements:** Draft detailed agreements outlining responsibilities, performance expectations, and payment terms. Include provisions for managing potential issues like stockouts or delivery delays to avoid conflicts and ensure smooth operations.
6. **Monitor and adjust performance:** Regularly review the VMI arrangement against the established metrics. Assess performance, identify areas for improvement, and make necessary adjustments to optimize the inventory management process and achieve desired outcomes.

Self Check 3

1. How can retailers ensure data security and confidentiality while sharing sensitive inventory and sales information with vendors in a VMI system?
2. What strategies can be implemented to effectively manage potential disruptions in VMI partnerships, such as vendor supply chain issues or unexpected changes in demand?

5. Summary

Chapter 4 explores key inventory replenishment strategies essential for optimizing supply chain operations. This chapter begins with JIT inventory management, a strategy designed to minimize inventory holding costs by aligning inventory levels closely with production schedules and demand. JIT emphasizes reducing waste and improving efficiency by ordering and receiving inventory only as needed for production or sales. This approach requires precise demand forecasting and reliable supplier relationships to avoid stockouts and disruptions. This chapter then delves into the EOQ model, a quantitative approach for determining the optimal order quantity that minimizes the total cost of inventory. EOQ balances ordering costs (expenses related to placing orders) and holding costs (costs of storing inventory) to find the most cost-effective order size. Assumptions of constant demand, lead time, and costs are integral to the EOQ model, though it may face limitations, such as not accommodating bulk discounts or varying demand patterns. Finally, this chapter addresses VMI, a collaborative approach where vendors assume responsibility for managing inventory levels at the retailer's location. Through VMI, vendors use real-time sales and inventory data to optimize stock levels, reduce excess inventory, and improve supply chain efficiency. Key practices include sharing accurate data, aligning processes, and fostering strong communication and partnerships. Together, these strategies (JIT, EOQ, and VMI) provide diverse methods for enhancing inventory management, reducing costs, and improving operational efficiency.

6. Case Study: Desll's Economic Order Quantity (EOQ) and Vendor Managed Inventory (VMI) Integration

Dell Technologies, a global leader in computer systems, has implemented sophisticated inventory management strategies to streamline its operations and maintain competitiveness. A critical component of Dell's strategy includes integrating Economic Order Quantity (EOQ) with Vendor Managed Inventory (VMI) practices. Dell employs the EOQ model to determine the optimal order size for components and raw materials. This model helps balance ordering costs and holding costs, ensuring that Dell maintains efficient inventory levels without incurring unnecessary expenses. The company uses the EOQ formula to calculate the ideal order quantity that minimizes the total inventory cost, taking into account factors such as demand rates, ordering costs, and holding costs.

In addition to EOQ, Dell utilizes Vendor Managed Inventory (VMI) for its supply chain operations. Through VMI, Dell's suppliers manage inventory levels at Dell's warehouses and manufacturing sites. This approach involves real-time data sharing. Dell provides its suppliers with real-time data on inventory levels and sales forecasts. This transparency allows suppliers to monitor stock levels and anticipate replenishment needs. Next is supplier responsibility where vendors are responsible for restocking inventory based on the data provided. They use this information to make informed decisions about order quantities and delivery schedules. Finally, Dell and its suppliers agree on performance metrics such as delivery reliability and inventory turnover rates to measure the success of the VMI arrangement. By using EOQ, Dell effectively manages inventory levels, minimizing both ordering and holding costs. Furthermore, VMI enhances stock availability and reduces the risk of stockouts, ensuring that Dell's production lines run smoothly without interruptions. Additionally, the integration of VMI fosters stronger relationships with suppliers, leading to better coordination and more reliable supply chain performance. However, EOQ and VMI integration requires careful coordination and system compatibility between Dell and its suppliers, which can be complex and time-consuming. Dell's combination of EOQ and VMI demonstrates the benefits of optimizing inventory management through strategic ordering and supplier collaboration. This integrated approach has helped Dell achieve cost savings, improve stock availability, and strengthen supplier partnerships.

Case Study Questions

1. How does Dell use EOQ model to balance ordering and holding costs, and what impact does this have on their overall inventory management?
2. What are the key benefits Dell has experienced from integrating VMI with its supply chain operations, particularly in terms of stock availability and supplier relationships?
3. What challenges might Dell face in integrating EOQ and VMI, and how can they address these challenges to ensure successful implementation and management of their inventory systems?

References

Edward, A.S., David F.P., and Douglas J.T. (2021). *Inventory and Production Management in Supply Chains,* 4th edn, Boca Raton, Florida, USA: CRC Press.

Steven, M.B. (2021). *Inventory Management*, 4th edn. Columbia, MD, USA: AccountingTools, Inc.

Vandeput, N. (2020). *Inventory Optimization: Models and Simulations,* 1st edn. Berlin, Germany: De Gruyter Publisher.

Chapter 5

Green Warehousing

Learning Outcome

By the end of this topic, you should be able to do the following:

1. Explain the importance of green warehousing.
2. Describe the strategy for going green.
3. Implement ISO 14000 for warehousing operations.

1. Introduction

The concept of green warehousing has gained significant importance in recent years as organizations strive to minimize their environmental footprint and enhance sustainability. Green warehousing involves the adoption of eco-friendly practices and technologies that reduce energy consumption, minimize waste, and promote the efficient use of resources. This chapter delves into the critical importance of green warehousing in modern supply chains and its role in fostering a sustainable future. By understanding the environmental impact of traditional warehousing operations, we can appreciate the need for greener alternatives. A comprehensive strategy for going green in warehousing encompasses several key elements, including energy-efficient lighting, renewable energy sources, sustainable building materials, and waste reduction initiatives. This chapter provides an in-depth look at these strategies, offering practical insights and best practices for warehouse managers and operators aiming to transition to greener operations. Additionally, the implementation of ISO 14000 standards plays a vital role in structuring organizational operations and ensuring environmentally responsible warehousing practices. ISO 14000 provides a framework for managing environmental aspects, compliance with regulations, and continuous improvement in environmental performance. This chapter will guide you through the steps necessary to integrate ISO 14000 into warehousing operations, ensuring a systematic approach to achieving and maintaining green objectives. By the end of this chapter, you will be equipped with the knowledge and tools to foster a more sustainable and environmentally conscious warehousing environment.

2. Importance of Green Warehousing

2.1 *Overview of green warehousing*

Green warehousing is an innovative approach to warehouse management that focuses on sustainability and reducing the environmental impact of warehousing operations. This concept encompasses a variety of practices and technologies aimed at minimizing energy consumption, reducing waste, and promoting the efficient use of resources. Green warehousing integrates sustainable practices into all aspects of warehousing operations, from construction and energy use to logistics and waste management. This approach not only benefits the environment but also improves efficiency, reduces costs, and enhances the overall reputation of the organization. Green warehousing offers numerous benefits that extend beyond environmental sustainability. By reducing greenhouse gas emissions, conserving energy and water, and minimizing waste, green warehousing significantly contributes to environmental preservation. These practices also lead to substantial cost savings, ensure regulatory compliance, enhance corporate image, and improve operational efficiency. As the demand for sustainable business practices continues to grow, the adoption of green warehousing is becoming increasingly essential for companies aiming to achieve long-term success and sustainability.

2.2 *Benefits of green warehousing*

Green warehousing involves adopting sustainable practices in warehouse operations to minimize environmental impact while improving efficiency and reducing costs. The benefits of green warehousing span various aspects, including environmental sustainability, cost savings, regulatory compliance, corporate image, and operational efficiency. Each of these areas is integral to creating a sustainable and economically viable warehousing system.

1. **Environmental impact:** One of the most significant benefits of green warehousing is the reduction in greenhouse gas emissions. Traditional warehousing operations often rely heavily on fossil fuels, contributing to a high carbon footprint. By incorporating renewable energy sources such as solar panels and wind turbines, warehouses can drastically reduce their reliance on non-renewable energy, thus lowering green house gas emissions. Additionally, energy-efficient systems like LED lighting and optimized HVAC (heating, ventilation, and air-conditioning) systems reduce overall energy consumption. Water conservation measures, such as rainwater harvesting and low-flow fixtures, further reduce the environmental impact by conserving a vital natural resource. Implementing robust recycling programs and waste management systems also minimizes waste generation, ensuring that less material ends up in landfills and more is recycled or reused.

2. **Cost savings:** Although the initial investment in green technologies and systems can be substantial, the long-term financial benefits are considerable. Energy-efficient systems and renewable energy sources reduce utility bills, resulting in significant cost savings over time. For instance, solar panels may have a high upfront cost, but they provide free, renewable energy for decades, lowering electricity costs significantly. Similarly, efficient water management practices reduce water bills. Waste reduction practices also contribute to cost savings by minimizing the costs associated with waste disposal and material procurement. By streamlining processes and reducing waste, warehouses can achieve higher levels of operational efficiency and lower operating expenses.

3. **Regulatory compliance:** As governments around the world implement stricter environmental regulations, businesses are under

increasing pressure to comply with these standards. Green warehousing helps companies stay ahead of regulatory requirements, ensuring compliance with environmental laws and avoiding potential fines. Adopting sustainable practices demonstrates a proactive approach to environmental responsibility, which can be particularly advantageous in industries with stringent regulatory frameworks. Compliance with environmental regulations not only avoids legal issues but also positions companies as leaders in sustainability, setting a positive example for the industry.

4. **Corporate image:** A commitment to sustainability can significantly enhance a company's corporate image. In today's market, consumers, investors, and stakeholders are increasingly valuing environmental responsibility. Companies that prioritize sustainability in their operations, including warehousing, are more likely to attract customers who prefer to support environmentally conscious businesses. A strong commitment to green practices can also appeal to investors looking for long-term, sustainable growth and employees seeking to work for companies that align with their values. This improved reputation can lead to increased customer loyalty, better investor relations, and a more motivated workforce.

5. **Operational efficiency:** Green warehousing practices contribute to overall operational efficiency. Efficient inventory management systems, such as just-in-time (JIT) inventory, reduce excess stock and waste, ensuring that products are available when needed without overstocking. Advanced demand forecasting minimizes the risk of holding unnecessary inventory, which reduces storage costs and waste. Energy management systems optimize energy use, reducing waste and ensuring that operations run smoothly and efficiently. Sustainable logistics practices, such as optimizing transportation routes and using eco-friendly vehicles, reduce fuel consumption and emissions while improving delivery times and reducing costs.

2.3 *Implementation considerations for green warehousing*

Implementing green warehousing practices requires careful planning and consideration to ensure success and sustainability. Key considerations include technology investment, employee engagement, and collaboration with stakeholders. The successful implementation of green warehousing requires a strategic approach that includes significant technology invest-ments, robust employee engagement, and active collaboration with stake-holders. By addressing these key considerations, organizations can effectively transition to sustainable warehousing practices, achieving long-term environmental and economic benefits. The effective adoption of environmentally sustainable practices in warehousing operations should consider the following:

1. **Technology investment:** The initial cost of implementing green technologies and systems can be substantial. Upgrading to energy-efficient lighting, installing renewable energy sources like solar panels, and integrating advanced energy management systems require significant capital investment. However, these investments offer sub-stantial long-term savings. Energy-efficient systems reduce utility bills, and renewable energy sources provide cost-free energy over time, offsetting the initial expenses. Moreover, these technologies significantly reduce the environmental footprint, contributing to sus-tainability goals. Organizations must conduct a thorough cost–benefit analysis to understand the financial and environmental returns on these investments. Long-term planning and budgeting for these upgrades can help manage costs effectively and ensure that the transi-tion to green warehousing is financially viable.

2. **Employee engagement:** Educating and training employees on the importance of sustainable practices is crucial for the successful imple-mentation of green warehousing initiatives. Employees at all levels need to understand the benefits of these practices and how they can contribute to environmental goals. Regular training sessions, work-shops, and informational materials can help build a culture of sustain-ability within the organization. Encouraging employees to participate in sustainability initiatives and providing incentives for green prac-tices can also enhance engagement. When employees are informed and motivated, they are more likely to adhere to sustainable practices, ensuring the long-term success of green warehousing efforts.

3. **Collaboration:** Collaboration with suppliers, logistics providers, and customers is essential to maximizing the impact of green warehous-ing. Working closely with suppliers to source sustainable materials, partnering with logistics providers to optimize transportation routes, and encouraging customers to participate in recycling programs can

significantly enhance the overall sustainability of the supply chain. Transparent communication and joint efforts to adopt sustainable practices ensure that all stakeholders are aligned with the organization's environmental goals. This collaborative approach not only amplifies the impact of green warehousing initiatives but also fosters strong relationships and shared commitment to sustainability across the supply chain.

Self Check 1

1. What are the main challenges companies face when transitioning to green warehousing and how can these challenges be effectively addressed?
2. How can companies enhance collaboration with suppliers, logistics providers, and customers to maximize the impact of green warehousing initiatives?

3. Strategy for Going Green

Implementing a green warehousing strategy involves a comprehensive approach that addresses energy efficiency, renewable energy, sustainable materials, waste management, water conservation, and logistics. By following these strategies, companies can effectively transition to green warehousing, reaping the benefits of reduced environmental impact, cost savings, regulatory compliance, improved corporate image, and enhanced operational efficiency.

1. **Energy efficiency:** One of the primary goals of green warehousing is to reduce energy usage. Energy efficiency is a cornerstone of green warehousing, focusing on reducing the amount of energy consumed in warehouse operations. This not only lowers operational costs but also minimizes the environmental impact by decreasing greenhouse gas emissions. This can be achieved through the installation of energy-efficient lighting systems, optimizing heating, ventilation, and air-conditioning (HVAC) systems, and utilizing natural light and skylights.

Energy-efficient lighting systems can be achieved using light-emitting diode (LED) lighting. LED lights are far more energy-efficient compared to traditional incandescent or fluorescent lights. LEDs consume less power, have a longer lifespan, and provide better illumination. This reduces both energy costs and the frequency of replacements. Moreover, installing motion sensors ensures that lights are only used when necessary. In areas of the warehouse that are not frequently occupied, lights automatically turn off when no movement is detected, further reducing energy consumption. Additionally, the use of smart lighting systems can be programmed to adjust the lighting intensity based on the time of day or specific operational needs. For instance, lighting can be dimmed during daylight hours when natural light is abundant.

Optimizing HVAC systems includes the use of efficient HVAC units. Modern HVAC systems are designed to be more energy-efficient, utilizing advanced technologies to provide heating, ventilation, and air-conditioning more effectively. Replacing older units with newer, energy-efficient models can lead to significant energy savings. Routine maintenance of HVAC systems ensures they operate at peak efficiency. This includes cleaning filters, checking for leaks, and ensuring all components are functioning correctly. Furthermore, zoning systems divide the warehouse into different sections that can be independently controlled. This means only areas in use are heated or cooled, reducing unnecessary energy consumption. Businesses can also utilize programmable thermostat controls that can be set to adjust temperatures based on occupancy and operational schedules. For example, temperatures can be reduced during non-operational hours or in less frequently used areas.

Utilizing natural light and skylights is another aspect of the energy efficiency in the warehouse. Daylighting by incorporating large windows, glass doors, and skylights allows natural light to penetrate the warehouse. This reduces the need for artificial lighting during daytime hours. Moreover, strategically placed skylights can provide ample natural light, especially in large warehouse spaces. They can be designed to minimize heat gain during summer and heat loss during winter, contributing to overall energy efficiency. Businesses can also use light tubes which are also known as solar tubes or light pipes as these devices capture daylight and deliver it to interior spaces. They are highly efficient and can significantly reduce the need for electric lighting. Using reflective materials on walls and ceilings can help maximize the distribution of natural light within the warehouse, enhancing the effectiveness of daylighting strategies. By focusing on energy efficiency, green warehousing not only promotes sustainability but also provides tangible economic benefits, making it a critical aspect of modern warehouse management.

2. **Renewable energy sources:** Incorporating renewable energy sources is a key component of green warehousing, aimed at reducing dependency on non-renewable energy and minimizing environmental impact. By harnessing clean, sustainable energy, warehouses can achieve significant cost savings and contribute to environmental sustainability. Incorporating renewable energy sources, such as solar panels, wind turbines, geothermal energy, and biomass energy, can help warehouses reduce their dependence on non-renewable energy. These systems not only lower energy costs but also reduce the carbon footprint of the warehouse.

Solar panels and photovoltaic (PV) systems convert sunlight directly into electricity. Installing PV panels on warehouse rooftops or adjacent land can generate a substantial amount of the energy required for operations. Furthermore, solar energy storage, coupling solar panels with battery storage systems, allows warehouses to store excess energy generated during sunny periods. This stored energy can be used during cloudy days or nighttime, ensuring a consistent power supply. Moreover, solar water heating where solar thermal systems can be used to heat water for various warehouse needs, such as cleaning processes and employee facilities, reduces the reliance on conventional water heaters.

Wind turbines are another alternative to renewable energy sources. On-site wind turbines can be installed on-site to generate electricity. Small to medium-sized wind turbines can produce sufficient power to support warehouse operations, especially in areas with high wind potential. Combining wind turbines with solar panels creates a hybrid renewable energy system. This ensures a more reliable and balanced energy supply by harnessing both solar and wind energy, which can be complementary (solar power during sunny days and wind power during windy conditions).

Geothermal energy using geothermal heat pump systems utilizes the stable temperatures below the earth's surface to heat and cool the warehouse. Geothermal heat pumps are highly efficient and can significantly reduce energy consumption for HVAC systems. Moreover, geothermal energy can be stored underground due to excess heat generated during the summer for use during the winter, optimizing energy usage throughout the year.

Biomass energy involves using organic materials such as wood pellets, agricultural residues, or other biological waste to generate heat. Biomass boilers can provide a sustainable heating solution for warehouses. Organic waste from warehouse operations, such as food or agricultural waste, can be processed in anaerobic digesters to produce biogas. This biogas can be used to generate electricity or heat, further reducing reliance on non-renewable energy sources. By integrating renewable energy sources into warehouse operations, companies can achieve substantial environmental and economic benefits, paving the way for a more sustainable and resilient future.

3. **Sustainable building materials:** Using sustainable building materials in the construction and maintenance of green warehouses is vital for reducing environmental impact and promoting sustainability. These materials are chosen for their lower environmental footprint, recyclability, and health benefits. The construction and maintenance of green warehouses often involve the use of sustainable and recycled building materials. This includes using low-VOC (volatile organic compounds) paints and adhesives, recycled steel, sustainably sourced wood, recycled and reclaimed materials, sustainable insulation materials, green roofing materials, and eco-friendly flooring.

Low-VOC paints and adhesives are preferred as VOCs are harmful chemicals found in many traditional paints and adhesives. They can diffuse into the air, contributing to indoor air pollution and health problems. Low-VOC paints and coatings release fewer VOCs, improving indoor air quality and reducing health risks for workers. Similar to paints, low-VOC adhesives are used for various construction and maintenance purposes. These adhesives are formulated to emit fewer toxic fumes, contributing to a healthier indoor environment.

Recycled steel is steel that has been reprocessed from scrap metal. Using recycled steel in construction reduces the need for virgin steel production, which is energy-intensive and generates significant greenhouse gas emissions. Recycling steel saves energy and natural resources. Steel is a highly durable material, providing long-lasting structural support. Recycled steel retains the same strength and performance characteristics as new steel, making it an excellent choice for sustainable construction.

Sustainably sourced wood is harvested from forests managed according to rigorous environmental and social standards. Certifications such as the Forest Stewardship Council (FSC) ensure that the wood comes from responsibly managed forests that prioritize biodiversity, conservation, and fair labor practices. Wood is a renewable resource and has a lower carbon footprint compared to many other building materials. Using sustainably sourced wood helps preserve forests, reduce deforestation, and support sustainable forestry practices.

Recycled and reclaimed materials include recycled concrete that can be recycled by crushing old concrete into aggregate, which can then be used in new construction projects. This reduces the demand for new raw materials and decreases landfill waste. Reclaimed wood and brick are salvaged from old buildings and repurposed for new construction. These materials add character and historical value to new buildings while reducing the need for new materials. Glass can be recycled and used in various construction applications, including windows, countertops, and flooring. Recycled glass reduces the need for new glass production and minimizes waste.

Sustainable insulation materials include cellulose insulation, which is made from recycled paper and cardboard. Cellulose insulation is an eco-friendly alternative to traditional insulation materials. It is effective in reducing energy consumption by providing excellent thermal performance. Next, sheep wool is a natural, renewable insulation material that is biodegradable and has good thermal and acoustic properties. It is also resistant to fire and pests.

Green roofing materials are cool roofing materials designed to reflect more sunlight and absorb less heat than standard roofing materials. This reduces the need for air-conditioning and lowers energy consumption. Green roofs are covered with vegetation and soil, providing natural insulation, reducing stormwater runoff, and improving air quality. They also create green spaces that support biodiversity and enhance the aesthetic appeal of warehouses.

Eco-friendly flooring includes bamboo flooring. Bamboo is a fast-growing, renewable resource that makes for a durable and attractive flooring material. It has a lower environmental impact compared to traditional hardwood flooring. Cork is harvested from the bark of cork oak trees without harming the trees. It is renewable, biodegradable, and provides excellent thermal and acoustic insulation as flooring. Incorporating sustainable building materials into green warehousing not only reduces the environmental impact but also creates healthier, more efficient, and cost-effective warehouse facilities. This approach aligns with broader sustainability goals and contributes to a more sustainable future.

4. **Waste reduction and recycling:** Effective waste management is a crucial aspect of green warehousing, aimed at minimizing environmental impact and promoting sustainability. By implementing robust waste reduction and recycling practices, warehouses can significantly reduce the amount of waste sent to landfills, conserve resources, and lower operational costs. The following are some of the key strategies for waste reduction and recycling in green warehousing:

Reducing packaging waste is possible using packaging designs that require less material without compromising product protection. This can include redesigning boxes, reducing the use of fillers, and eliminating unnecessary packaging layers. Furthermore, shipping products in bulk rather than individually packaged items can significantly reduce the amount of packaging waste. Additionally, implementing reusable packaging solutions such as durable plastic crates, containers, and pallets that can be used multiple times before being recycled helps reduce packaging waste.

Recycling materials is possible through recycling programs. Companies can establish comprehensive recycling programs to segregate and recycle materials such as paper, cardboard, plastics, metals, and glass. This involves setting up designated recycling bins and ensuring that employees are trained on proper recycling procedures. Companies may also use cardboard compactors to efficiently manage and store cardboard waste before it is sent for recycling. This reduces the volume of waste and transportation costs. Implementing systems to collect and recycle plastic packaging, shrink wrap, and other plastic materials used in warehousing operations is another strategy.

Proper disposal of hazardous substances is done by identifying and properly disposing hazardous substances such as chemicals, batteries, and electronic waste. This involves following regulatory guidelines and using certified disposal services to prevent environmental contamination. Companies may also store hazardous materials in secure, labeled containers to prevent leaks and spills that could harm the environment and human health.

Composting organic waste involves segregating food scraps, yard waste, and biodegradable materials separately from other types of waste. It is also important to set up on-site composting systems to convert organic waste into nutrient-rich compost. This compost can be used for landscaping around the warehouse or distributed to local farms and gardens.

Reusing pallets and containers is done via pallet pooling programs where pallets are shared among multiple companies, reducing the need for single-use pallets and promoting reuse. Moreover, damaged pallets and containers can be repaired and reused instead of discarding them. Companies can implement a system to regularly inspect and maintain pallets and containers to extend their lifespan.

Implementing circular economy practices through product life cycle management is helpful. Companies should design products and packaging with their entire life cycle in mind, focusing on reuse, recycling, and recovery of materials. Furthermore, companies can work with suppliers (supplier collaboration) to develop and implement circular economy practices, such as returning packaging materials for reuse or recycling.

5. **Water conservation:** Water conservation is a crucial aspect of green warehousing, aimed at reducing water usage, preserving this vital resource, and minimizing the environmental impact of warehouse operations. Efficient water management practices not only contribute to sustainability but also lower operational costs. Key water conservation strategies in green warehousing include the following:

Rainwater harvesting involves collecting and storing rainwater from rooftops and other surfaces. This water can be captured in large tanks or cisterns for later use. Harvested rainwater can be used for non-potable purposes such as landscape irrigation, toilet flushing, cleaning, and cooling systems. This reduces the demand for municipal water and conserves potable water resources. Depending on its intended use, harvested rainwater may need to be filtered and treated to remove debris and contaminants. Simple filtration systems can make the water suitable for most non-potable applications.

Water recycling is done through greywater systems that collect and treat wastewater from sinks, showers, and other non-toilet plumbing fixtures. This treated greywater can be reused for irrigation, flushing toilets, and other non-potable purposes. Water used in various warehouse processes, such as cooling systems or equipment washing, can be treated and recycled. Implementing closed-loop systems ensures water is reused multiple times before being discharged. In some cases, advanced water treatment technologies such as reverse osmosis or ultraviolet (UV) purification can be employed to recycle water for more sensitive applications.

Low-flow fixtures and urinals significantly reduce the amount of water used per flush. Dual-flush toilets offer the option of using less water for liquid waste and more for solid waste. Low-flow faucets and showerheads use aerators and other technologies to reduce water flow without compromising performance. This helps in conserving water in restrooms and employee shower areas. Sensor-activated faucets and flush valves minimize water wastage by ensuring that water only flows when needed. These fixtures are especially effective in high-traffic areas.

Efficient landscaping using native or drought-tolerant plants reduces the need for irrigation. These plants are adapted to local climate conditions and require minimal water. Drip irrigation delivers water directly to the root zone of plants, minimizing evaporation and runoff. This method is highly efficient and reduces water usage compared to traditional sprinkler systems. Furthermore, applying mulch around plants helps retain soil moisture, reduce evaporation, and suppress weeds, further conserving water.

Leak detection and repair while conducting regular inspections of plumbing systems and fixtures can identify and repair leaks promptly. Even small leaks can lead to significant water wastage over time. Companies may also install automated water monitoring systems that detect leaks and unusual water usage patterns. These systems can alert maintenance personnel to potential issues, ensuring quick resolution.

Employee awareness and training should include water conservation education. Educating employees about and providing training on the importance of water conservation are best practices. Encouraging water-saving behaviors, such as reporting leaks and using water efficiently, is also important. Furthermore, companies should place signage and reminders in restrooms, break rooms, and other areas to promote water conservation. Simple reminders can have a significant impact on reducing water usage.

6. **Green logistics:** Green logistics involves the integration of environmentally friendly practices into the logistics and supply chain management of warehouses. This approach aims to reduce the environmental impact of warehousing operations while improving efficiency and cost effectiveness. Key green logistics strategies in green warehousing include the following:

Optimizing transportation routes can be done using route planning software. Utilizing advanced route planning software helps optimize delivery routes to minimize travel distances, reduce fuel consumption, and decrease emissions. These tools can analyze traffic patterns, road conditions, and delivery schedules to find the most efficient routes. Companies may also combine multiple shipments into a single trip, reducing the number of trips needed, thereby saving fuel and reducing emissions. This practice is particularly effective for deliveries to locations in close proximity to one another. Furthermore, ensuring that vehicles are fully loaded to their optimal capacity reduces the number of trips required and maximizes the efficiency of each journey. Proper load planning can significantly cut down on fuel usage and costs.

Using eco-friendly vehicles means utilizing electric and hybrid vehicles. Switching to electric or hybrid vehicles for transportation and delivery can substantially reduce greenhouse gas emissions and reliance on fossil fuels. These vehicles are particularly suitable for short-range deliveries and urban logistics. Additionally, using alternative fuels such as biodiesel, natural gas, or hydrogen can lower the carbon footprint of transportation operations. These fuels produce fewer emissions compared to traditional diesel or gasoline. Companies should also conduct regular maintenance of vehicles to ensure they operate efficiently and produce fewer emissions. This includes checking tire pressure, engine tuning, and ensuring vehicles are in good working condition.

Efficient inventory management systems include demand forecasting. Implementing advanced demand forecasting techniques helps predict customer demand more accurately, reducing the need for excess inventory. This minimizes waste and lowers the resources required for storage and handling. Adopting just-in-time (JIT) inventory systems ensures that products are delivered and stocked only as needed, reducing the amount of inventory held in the warehouse. This decreases storage costs and minimizes the risk of overstocking and obsolescence. Similarly, using automated systems for inventory tracking and management enhances accuracy and efficiency. Technologies like RFID, barcoding, and inventory management software streamline processes, reducing errors and waste.

Reducing excess stock and waste by monitoring and improving inventory turnover rates helps ensure that stock is rotated efficiently, reducing the risk of products becoming obsolete or expiring. This practice minimizes waste and improves cash flow. Moreover, implementing sustainable packaging solutions, such as recyclable or biodegradable materials, reduces packaging waste. Optimizing packaging sizes to fit products more closely also reduces material usage and waste. Likewise, establishing reverse logistics systems for the return, refurbishment, and recycling of products minimizes waste and promotes the reuse of materials. This includes handling returns efficiently and ensuring that products are processed for resale or recycling.

Energy-efficient warehousing operations can be guaranteed through green building certifications. Companies may design warehouses to meet green building standards, including LEED (Leadership in Energy and Environmental Design), which ensures energy-efficient operations and low environmental impact. Through it, incorporating renewable energy sources such as solar panels or wind turbines to power warehouse operations reduces reliance on non-renewable energy and lowers emissions. Also, implementing energy management systems to monitor and optimize energy usage within the warehouse is important. This includes using smart sensors, automated controls, and energy-efficient lighting and HVAC systems.

Self Check 2

1. How can companies effectively educate and motivate employees to adopt and maintain sustainable practices within the warehouse?

2. Based on your experience, what are the first steps a company should take when beginning to implement a green warehousing strategy?

4. Implementing ISO 14000 for Warehousing Operations

ISO 14000 is a family of standards related to environmental management that helps organizations minimize their environmental impact. Implementing ISO 14000 in warehousing operations involves several steps to ensure compliance and continuous improvement. Implementing ISO 14000 for warehousing operations requires commitment, planning, and continuous effort as shown in Figure 1.

Through the following steps, warehouses can achieve significant environmental and operational benefits, contributing to broader sustainability goals:

1. **Understand ISO 14000 requirements:** Learn the specific requirements and guidelines of ISO 14001, the key standard in the ISO 14000 family, focused on environmental management systems (EMSs). This involves understanding the framework and principles of ISO 14001. The company should also conduct a gap analysis to compare current warehousing practices with ISO 14001 standards.

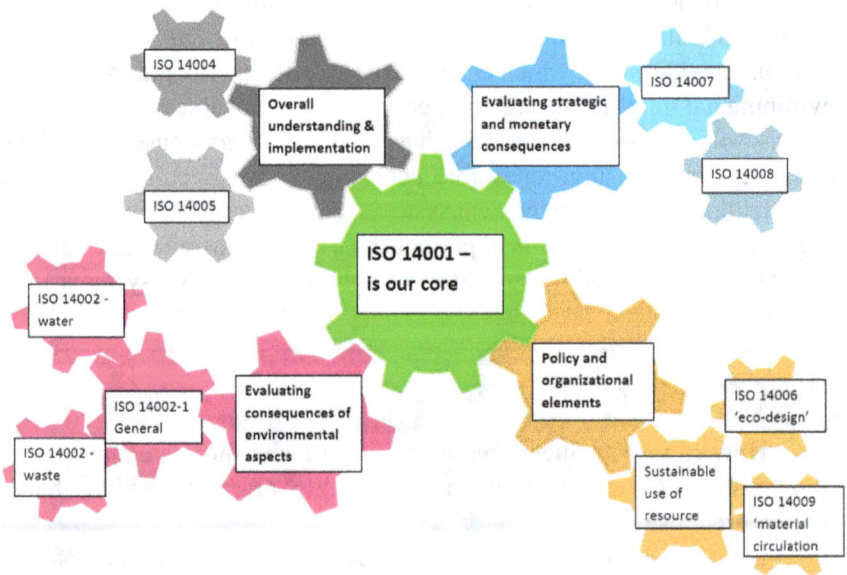

Figure 1.   ISO 14000 framework.

Identify discrepancies and areas needing improvement. This helps in pinpointing specific changes required to align the warehouse's environmental management practices with ISO 14001 requirements.

2. **Develop an environmental policy:** Draft a policy that clearly states the warehouse's dedication to sustainability, legal compliance, and ongoing improvement. This statement should be concise and reflect the core values of the organization. Furthermore, establish specific, measurable environmental objectives and targets derived from the policy. Focus on critical areas like reducing energy consumption, improving waste management, and conserving resources. These goals should be realistic, time-bound, and designed to drive continuous environmental performance improvement.

3. **Plan the implementation:** Create a comprehensive action plan that details the specific steps required to achieve the set environmental objectives. This plan should include timelines, assigned responsibilities, and the necessary resources for each task. Form an EMS team dedicated to implementing and maintaining the environmental management system. Ensure that team members are equipped with the necessary skills and training to effectively execute their roles. Clear role assignment and adequate training are crucial for the successful adoption and ongoing management of the EMS.

4. **Implement the EMS:** Develop and maintain comprehensive documentation of all procedures, processes, and records in line with ISO 14001 requirements, including environmental policies, risk assessments, and operational controls. Educate all employees on their specific roles and responsibilities within the EMS. Ensure they fully understand the environmental policy and the environmental impact of their actions. Companies should also establish and enforce operational controls to manage significant environmental aspects. This includes protocols for handling hazardous materials, waste management, and energy-saving measures to mitigate environmental impact effectively.

5. **Monitor and measure:** Continuously gather and analyze data on the warehouse's environmental performance. Monitor key performance indicators (KPIs) such as energy consumption, waste generation, and water usage to track progress toward environmental goals. It is also important to perform regular internal audits to evaluate compliance with the environmental management system (EMS). Identify areas for improvement and ensure adherence to ISO 14001 standards. Use

audit findings to refine strategies and practices, driving continuous enhancement of environmental performance and operational efficiency.

6. **Review and improve:** Conduct regular management review meetings to assess the effectiveness of the EMS. Review audit results, performance data, and progress toward environmental objectives to ensure alignment with goals and standards. The company may apply the Plan-Do-Check-Act (PDCA) cycle to foster the ongoing improvement of the EMS. Implement corrective actions to address any nonconformities and update procedures based on new insights, feedback, and advancements in technology. This iterative approach ensures that the EMS evolves and adapts to enhance environmental performance and compliance continually.

7. **Certification:** Select an accredited certification body with experience in ISO 14001 to conduct an external audit. Verify their credentials and expertise to ensure a thorough evaluation of your EMS. Prepare thoroughly for the external audit by updating all relevant documentation and ensuring that employees are knowledgeable and ready to demonstrate compliance with the EMS. Successfully pass the external audit to obtain ISO 14001 certification. Display the certification prominently to showcase your commitment to effective environmental management to stakeholders and enhance your organization's credibility.

4.1 *Benefits of ISO 14000 implementation*

Obtaining ISO 14000 certification can be considered a sign of commitment to the environment, which can be used as a marketing tool for companies. It may also help companies meet environmental regulations that are imposed by governments of areas in which they do business. Some of the benefits include the following:

1. **Enhanced environmental performance:** Systematic management of environmental aspects improves efficiency and reduces impact. For example, by implementing energy-efficient lighting and optimizing HVAC systems, a warehouse can significantly cut energy consumption. Additionally, proper waste segregation and recycling practices reduce the amount of waste sent to landfills. This not only minimizes pollution but also lowers operational costs. These practices

collectively enhance the environmental performance of the warehouse, demonstrating a commitment to sustainability while achieving tangible operational benefits.

2. **Regulatory compliance:** Implementing a robust EMS helps ensure compliance with environmental regulations. For example, adhering to ISO 14001 standards involves regular monitoring and reporting of emissions, which aligns with legal requirements. By systematically managing waste and hazardous materials, a warehouse can avoid fines for improper disposal or violations of regulations. This proactive approach reduces the risk of legal issues and demonstrates a commitment to meeting environmental laws and standards.

3. **Cost savings:** Improving efficiency and reducing resource use can lead to significant cost savings. For example, by installing energy-efficient LED lighting and optimizing HVAC systems, a warehouse can lower its energy bills. Implementing waste reduction practices, such as recycling and minimizing packaging, reduces disposal costs and material expenses. Additionally, conserving water through low-flow fixtures decreases utility costs. These measures collectively reduce operational expenses, making the warehouse more cost-effective while also supporting sustainability goals.

4. **Market advantage:** Obtaining ISO 14001 certification highlights a warehouse's commitment to sustainability, providing a competitive edge. For example, a certified warehouse can leverage its green credentials to attract environmentally conscious clients and partners, enhancing its market reputation. This certification can also differentiate the warehouse from competitors, making it a preferred choice for businesses seeking to align with sustainable practices. The visible commitment to environmental management can build trust and credibility, potentially leading to increased business opportunities and market share.

Self Check 3

1. What are the common challenges organizations face when implementing ISO 14000 standards?
2. How can organizations measure the tangible benefits of ISO 14000 certification?

5. Summary

Green warehousing is essential for advancing environmental sustainability in logistics and supply chain operations. By focusing on reducing energy consumption, minimizing waste, and lowering greenhouse gas emissions, green warehousing not only contributes to environmental protection but also enhances operational efficiency and cost savings. Key strategies for transitioning to green warehousing include adopting energy-efficient systems, utilizing renewable energy sources, and incorporating sustainable building materials. Energy efficiency can be achieved through upgrades such as LED lighting and optimized HVAC systems, while renewable energy sources like solar panels can reduce dependence on non-renewable power. Sustainable materials, including recycled and low-VOC products, further support environmental goals. Implementing ISO 14000 standards, specifically ISO 14001, provides a structured approach to managing environmental performance. The ISO 14000 certification process begins with understanding the standards requirements and conducting a gap analysis to identify areas for improvement. Developing an environmental policy that outlines commitment, objectives, and targets is crucial. Planning involves creating an action plan and assigning roles to an Environmental Management System (EMS) team. The EMS should be implemented through comprehensive documentation, employee training, and operational controls. Monitoring and measuring involve collecting data on key performance indicators, conducting internal audits, and using the Plan-Do-Check-Act (PDCA) cycle for continuous improvement. Certification by an accredited body verifies compliance and showcases commitment to environmental management. Overall, green warehousing and ISO 14000 certification enhance environmental performance, offer cost savings, ensure regulatory compliance, and provide a market advantage through demonstrated sustainability.

6. Case Study: EcoLogistics Inc.

EcoLogistics Inc., a leading logistics provider with a network of warehouses across North America, embarked on a journey to implement green warehousing practices. The company's goal was to enhance its environmental performance, reduce operational costs, and comply with increasing regulatory requirements. EcoLogistics implemented several green strategies including energy efficiency. The company upgraded its warehouses with LED lighting, reducing energy consumption by 40%. Advanced HVAC systems with programmable thermostats were installed, cutting heating and cooling costs by 25%. Furthermore, the company also implemented renewable energy with solar panels being installed on the rooftops of major warehouses. This initiative provided approximately 30% of the total energy needs, significantly lowering reliance on non-renewable energy sources. EcoLogistics also used sustainable materials. During renovations, EcoLogistics used recycled steel and low-VOC paints. These materials minimized environmental impact and improved indoor air quality. Moreover, comprehensive recycling programs were introduced, reducing landfill waste by 50%. The company also implemented a pallet reuse program, further decreasing waste. EcoLogistics Inc. also focused on water conservation. A low-flow fixture and a rainwater harvesting system were installed, reducing water consumption by 35%.

EcoLogistics adopted ISO 14001 to formalize its environmental management system through gap analysis. The company assessed existing practices against ISO 14001 requirements, identifying key areas for improvement. An environmental policy was also developed, including clear objectives for energy use, waste management, and resource conservation. Next, a detailed action plan was created, assigning responsibilities and setting timelines for achieving environmental targets. EcoLogistics also focused on employee training. Staff received training on the new environmental procedures and their roles in achieving the company's sustainability goals. Furthermore, regular audits and data collection helped track progress. After a successful external audit, EcoLogistics received ISO 14001 certification, underscoring its commitment to environmental management. As a result, EcoLogistics realized significant benefits, including a 20% reduction in operational costs due to energy savings and improved waste management. The company enhanced its market reputation and achieved compliance with environmental regulations, demonstrating its commitment to sustainability and gaining a competitive edge in the logistics industry.

Case Study Questions

1. What were the key strategies EcoLogistics Inc. implemented to achieve energy efficiency and reduce operational costs in its green warehousing practices?
2. How did EcoLogistics Inc. use ISO 14001 to formalize its environmental management system, and what were the outcomes of achieving this certification?
3. What were the main challenges EcoLogistics Inc. faced during the implementation of green warehousing practices and ISO 14001 certification, and how did the company address these challenges?

Further Reading

Bak, O. (2021). *Sustainable and Green Supply Chains and Logistics Case Study Collection*, 1st edn. New York: Kogan Page.

Frazelle, E.H. (2023). *World-Class Warehousing and Material Handling*, 2nd edn. London, UK: McGraw Hill.

McKinnon, A., Browne, M., Whiteing, A., and Piecyk, M. (2016). *Green Logistics: Improving the Environmental Sustainability of Logistics*, 3rd edn. London, UK: Kogan Page.

Chapter 6

Warehouse Management Systems

1. Introduction

Warehouse management systems (WMSs) are essential tools in modern supply chain management. These systems facilitate the management of warehouse operations, from inventory tracking to order fulfillment. The importance of WMSs cannot be overstated, as they play a critical role in ensuring the efficiency and accuracy of warehouse processes. Historically, warehouse management was a manual and labor-intensive process. However, with the advent of technology, WMSs have evolved significantly, integrating sophisticated software solutions to automate and optimize warehouse operations.

The historical background of WMS dates back to the early days of warehousing when simple manual methods were used to keep track of inventory. Over time, the need for more efficient and accurate systems led to the development of early computerized inventory systems. Key milestones in the evolution of WMS include the introduction of barcode technology in the 1970s, the adoption of radio frequency identification (RFID) in the 1990s, and the integration of WMS with enterprise resource planning (ERP) systems in the early 2000s.

2. Components of WMS

A comprehensive WMS comprises several critical components that work together to ensure seamless warehouse operations. One of the primary components is inventory tracking. This involves using technologies such as barcodes and RFID to maintain real-time visibility of inventory levels. Real-time updates help in reducing discrepancies and ensuring accurate inventory counts. Cycle counting and reconciliation are also included in inventory tracking, which helps maintain inventory accuracy and reduce shrinkage.

Order management is another important component. This encompasses the entire process of handling customer orders, from order receipt to fulfillment. Order management systems are integrated with ERP and Customer Relationship Management (CRM) systems to ensure smooth data flow and efficient processing. Backorder management, which handles situations where inventory is insufficient to fulfill orders, is also an essential aspect of order management.

Labor management is another critical component of WMS. It involves workforce planning, scheduling, and productivity tracking. By monitoring employee performance and productivity, warehouse managers can optimize labor allocation and implement incentive programs to improve efficiency. Detailed reporting on labor performance helps identify areas for improvement and ensure that the workforce is effectively utilized.

Yard and dock management is crucial for managing the movement of goods in and out of the warehouse. This involves optimizing the yard layout for efficient loading and unloading, scheduling dock appointments to avoid congestion, and integrating with Transportation Management Systems (TMSs) to ensure seamless coordination between warehouse and transportation operations.

3. Key Functions in WMS

Receiving: Warehouse management systems streamline the receiving process by automating the capture of incoming goods data. When products arrive at the warehouse, WMS uses barcode scanning or RFID technology to log each item into the system, ensuring real-time updates and immediate visibility of new inventory. This reduces errors, accelerates the receiving process, and ensures that the inventory is accurately recorded. The system can also compare received items against purchase orders to identify discrepancies and initiate returns if necessary (Figure 1).

Putaway: After receiving, the WMS directs the putaway process, optimizing the placement of goods within the warehouse. The system uses

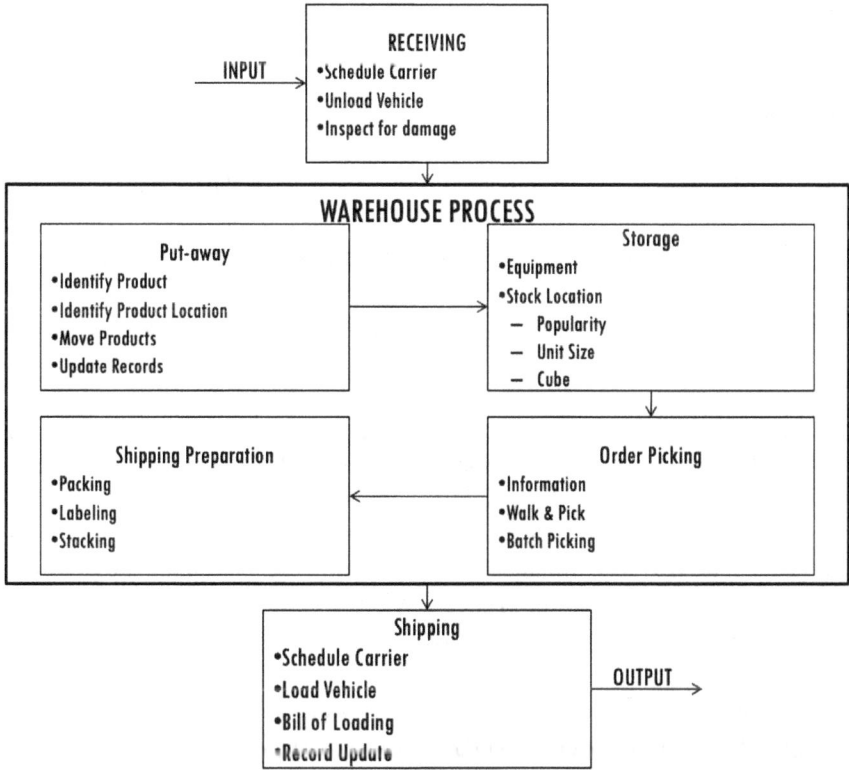

Figure 1.   Key processes in warehouse management.

algorithms to determine the most efficient storage location based on factors such as item size, weight, demand frequency, and proximity to picking areas. By guiding workers to the optimal storage locations, WMS helps maximize space utilization and ensures that items are easy to retrieve later. This reduces travel time within the warehouse and improves overall operational efficiency.

Picking: Picking is one of the most labor-intensive processes in a warehouse, and WMS significantly enhances its efficiency. The system generates pick lists and provides instructions to warehouse workers, often through mobile devices or voice-directed picking systems. WMS supports various picking methods, including wave picking, batch picking, and zone picking, depending on the warehouse layout and order characteristics. By optimizing the picking route and method, WMS reduces the time and effort required to gather items for orders, thus improving productivity and accuracy.

Staging: Staging involves preparing picked-up items for shipping. WMS coordinates the staging process by designating specific areas where items are consolidated before being packed and shipped. The system ensures that items are organized and grouped according to their respective orders, reducing the chances of errors. It also provides real-time updates on the status of orders, allowing warehouse managers to monitor progress and make adjustments as needed. WMS's efficient staging ensures smooth and accurate order fulfillment.

Shipping: In the shipping phase, WMS generates shipping labels, packing lists, and necessary documentation for each order. The system coordinates with carriers and schedules pickups, ensuring that shipments are dispatched on time. WMS can also track shipments and provide customers with real-time updates, enhancing transparency and customer satisfaction. By automating the shipping process, WMS minimizes errors and delays, ensuring that orders are delivered accurately and on schedule.

Cross-docking: WMS supports cross-docking, a method in which incoming goods are transferred directly to outbound shipping without being stored. This is particularly useful for high-turnover items. The system coordinates the timing and movement of goods, ensuring seamless transitions between receiving and shipping areas. Cross-docking reduces handling and storage costs, accelerates order fulfillment, and minimizes inventory holding times.

Cycle counting:
To maintain inventory accuracy, WMS facilitates cycle counting, a method of counting a portion of inventory on a regular basis. The system schedules cycle counts and generates lists of items to be counted, ensuring that all inventory is periodically verified. This proactive approach helps identify and resolve discrepancies before they become significant issues, maintaining high levels of inventory accuracy and reducing the need for disruptive full-scale physical counts.

Yard and dock management:
WMS also assists in yard and dock management by scheduling and tracking the movement of trailers and trucks within the yard. The system optimizes dock appointments and ensures that loading and unloading operations are managed efficiently. This reduces congestion, minimizes waiting times, and enhances the coordination between warehouse operations and transportation.

Labor management:
WMS includes labor management functionalities that track employee performance, manage work assignments, and optimize labor allocation. The system monitors key performance indicators (KPIs) such as picking rates, accuracy, and productivity, providing insights that help managers make informed decisions about workforce planning and training.

Reporting and analytics:
A key advantage of WMS is its robust reporting and analytics capabilities. The system collects data on all warehouse activities, generating detailed reports and dashboards that provide insights into operational performance. Managers can use this data to identify trends, uncover inefficiencies, and make data-driven decisions to improve warehouse operations. In fact, most of the improvement should be targeted at the picking process, which contributes to 55% of operating costs (Figure 2).

Benefits of implementing WMS:
Implementing a WMS offers numerous benefits, including improved inventory accuracy, enhanced customer service, increased efficiency and

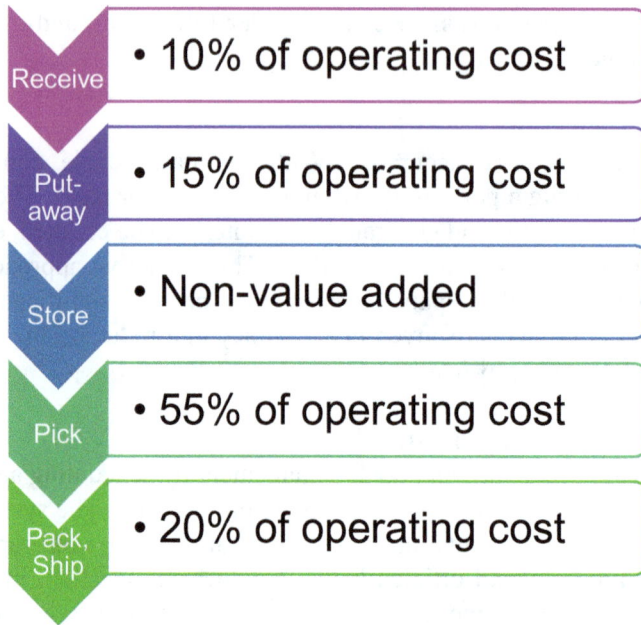

Figure 2. Breakdown of operating cost by process.

productivity, better space utilization, and data-driven decision-making. Improved inventory accuracy ensures that the right products are available when needed, reducing stockouts and overstock situations. Enhanced customer service results from faster and more accurate order fulfillment, leading to higher customer satisfaction.

Increased efficiency and productivity are achieved through automation and optimization of warehouse processes leading to increased efficiency and productivity. By streamlining operations, WMS reduces manual labor and minimizes errors. Better space utilization is another benefit, as WMS optimizes storage locations and ensures efficient use of warehouse space. Data-driven decision-making is facilitated by the detailed reporting and analytics capabilities of WMS, providing valuable insights into warehouse operations and helping managers make informed decisions.

4. Types of WMSs

There are several types of WMSs available, each with its own advantages and disadvantages. Standalone WMSs are dedicated systems that offer comprehensive warehouse management capabilities. These systems are highly customizable and can be tailored to the specific needs of the warehouse. ERP modules, on the other hand, are integrated with broader ERP systems and provide basic warehouse management functionalities. They are suitable for organizations that require a unified system for managing all aspects of their operations. Cloud-based WMSs are becoming increasingly popular due to their flexibility and scalability. These systems are hosted on remote servers and accessed via the Internet, allowing for easy updates and integration with other cloud-based applications. On-premises WMSs are installed and operated on the organization's own servers, offering greater control and customization but requiring significant upfront investment and ongoing maintenance.

Examples of WMSs include the following:

SAP Extended Warehouse Management (EWM)
- It offers advanced capabilities for inventory management and optimized warehouse processes.
- It is integrated with SAP's broader ERP system for seamless data flow and management.

Oracle Warehouse Management Cloud
- It is a robust, cloud-based WMS solution with features such as mobile barcoding and inventory optimization.
- It is scalable and designed to integrate with other Oracle applications.

Manhattan Associates WMS
- It is known for its comprehensive suite of features, including labor management, slotting optimization, and real-time data visibility.
- It is widely used in retail, manufacturing, and distribution industries.

Infor CloudSuite WMS
- It is a cloud-based WMS offering advanced inventory management, labor management, and 3PL billing.
- It is part of Infor's comprehensive suite of enterprise solutions.

Blue Yonder (formerly JDA) WMS
- It provides end-to-end visibility and control over warehouse operations.
- It focuses on AI and ML capabilities to drive automation and efficiency.

HighJump (now part of Körber)

- It is known for its adaptability and scalability, it offers features such as voice-directed warehousing and advanced analytics.
- It is suitable for various industries, including e-commerce, retail, and third-party logistics.

These WMS solutions are designed to address different aspects of warehouse management and can be tailored to fit the specific needs of various industries. They offer a range of functionalities, from basic inventory tracking to sophisticated, AI-driven warehouse optimization.

5. Implementation Process

Implementing a WMS involves several steps, starting with a needs assessment to determine the specific requirements of the warehouse, as shown in Figure 3.

This is followed by system selection, where different WMS options are evaluated based on their features, cost, and compatibility with existing systems. Customization and configuration are then carried out to tailor the system to the warehouse's specific needs. This involves setting up workflows, defining user roles, and integrating with other systems. Testing and training are critical to ensuring that the WMS functions correctly and that users are familiar with its features. This entails thoroughly testing all

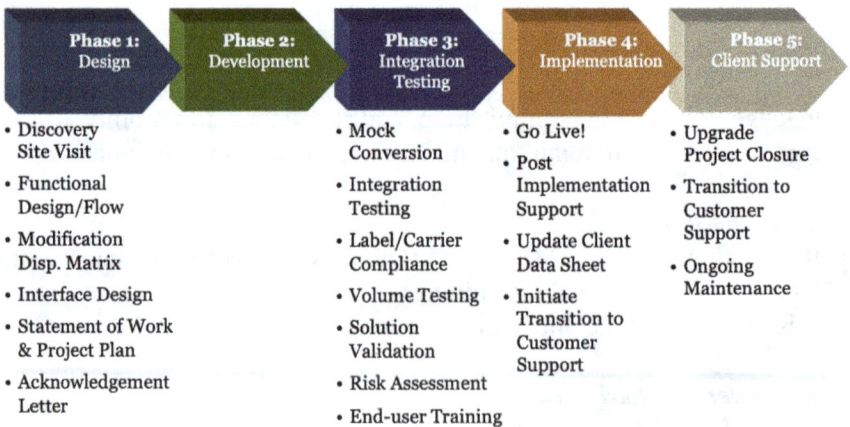

Phase 1: Design	Phase 2: Development	Phase 3: Integration Testing	Phase 4: Implementation	Phase 5: Client Support
• Discovery Site Visit		• Mock Conversion	• Go Live!	• Upgrade Project Closure
• Functional Design/Flow		• Integration Testing	• Post Implementation Support	• Transition to Customer Support
• Modification Disp. Matrix		• Label/Carrier Compliance	• Update Client Data Sheet	• Ongoing Maintenance
• Interface Design		• Volume Testing	• Initiate Transition to Customer Support	
• Statement of Work & Project Plan		• Solution Validation		
• Acknowledgement Letter		• Risk Assessment		
		• End-user Training		

Figure 3.   Implementation of WMS.

functionalities and providing comprehensive training to all users. The go-live phase entails deploying the system in the real world and providing ongoing support to address any issues that arise. Post-implementation support is essential to ensure the system continues to operate smoothly and to address any challenges that may arise.

Implementing a WMS can be challenging, with common issues including data migration, integration with existing systems, user resistance, and cost considerations. Data migration involves transferring data from legacy systems to the new WMS, which can be complex and time-consuming. Ensuring data accuracy and completeness is critical to avoid issues post-implementation. Integration with existing systems, such as ERP, CRM, and TMS, is another challenge. Ensuring seamless data flow between systems is essential for efficient operations. User resistance and training are also significant challenges, as employees may be resistant to change and require extensive training to become proficient with the new system. Finally, cost and ROI considerations are important, as implementing a WMS involves significant investment. It is essential to conduct a thorough cost–benefit analysis to ensure that the benefits justify the costs.

6. Future Trends in WMS

The future of WMS is shaped by emerging technologies and trends. Integration with the Internet of Things (IoT) and automation is becoming increasingly prevalent, with smart sensors and automated systems enhancing warehouse operations. Artificial intelligence (AI) and machine learning (ML) are being used to optimize processes and predict demand, leading to more efficient and accurate operations. Enhanced analytics and reporting capabilities are also shaping the future of WMS, providing deeper insights into warehouse operations and enabling data-driven decision-making. Sustainability and green logistics are becoming more important, with WMS playing a key role in optimizing processes to reduce environmental impact.

7. Case Study

Implementing a Warehouse Management System (WMS) in a Philippine Distribution Company.

Background

Company profile: ABC Distributors, Inc. is a mid-sized distribution company in the Philippines specializing in fast-moving consumer goods (FMCG). Operating a 10,000-square-meter warehouse in Laguna, the company supplies products to over 500 retail outlets nationwide.

Challenges:

1. **Manual processes:** Inventory management relied on spreadsheets, leading to frequent stock discrepancies.
2. **Order fulfillment errors:** Picking errors averaged 5%, resulting in customer complaints and returns.
3. **High operating costs:** Labor-intensive workflows increased operational expenses.
4. **Scalability issues:** The company struggled to meet growing demand due to inefficient systems.

To address these issues and prepare for future growth, ABC Distributors decided to implement a Warehouse Management System (WMS).

Objective

To optimize warehouse operations through the implementation of a WMS, aiming to improve inventory accuracy, reduce order errors, and enhance overall efficiency.

Implementation process

1. **Needs assessment:**
   A third-party consultant conducted an operational audit and identified key pain points, including the following:
   - Long order processing times (average of 3 hours per order).
   - Inefficient use of warehouse space.
   - Poor real-time visibility of inventory.
2. **Vendor selection:**
   After evaluating several local and international WMS providers, ABC Distributors chose a cloud-based WMS tailored for small and medium

enterprises (SMEs) in the Philippines. The selected WMS offered features such as the following:

- Barcode scanning for inventory tracking.
- Real-time dashboards for stock levels and order status.
- Integration with ABC's existing Enterprise Resource Planning (ERP) system.

3. **System implementation:**
   - **Phase 1:** Data migration and system setup. All existing inventory data was digitized and uploaded to the WMS.
   - **Phase 2:** Hardware installation, including barcode scanners and mobile devices for warehouse staff.
   - **Phase 3:** Staff training and pilot testing in one section of the warehouse.
   - **Phase 4:** Full rollout across the entire warehouse, supported by on-site vendor representatives.

4. **Change management:**
   To address employee resistance, management conducted workshops to demonstrate the WMS's benefits and how it would simplify daily tasks. Key warehouse staff were appointed as "change champions" to guide peers during the transition.

Results

1. **Improved efficiency:**
   - Order processing time reduced from 3 hours to 1 hour.
   - Picking and packing times decreased by 40%.

2. **Enhanced accuracy:**
   - Inventory accuracy increased from 85% to 99.5%.
   - Picking errors dropped from 5% to less than 0.5%.

3. **Cost savings:**
   - Labor costs reduced by 20% as manual processes were replaced by automation.
   - Warehouse space utilization improved by 30%, delaying the need for expansion.

4. **Better customer satisfaction:**
   - Faster delivery times and fewer errors led to a 15% increase in customer satisfaction scores.

5. **Scalability:**
   - The WMS enabled the company to handle a 25% increase in order volume without adding significant headcount.

Key learnings:
1. **Local adaptations:**
   The WMS was configured to address local requirements, such as integration with the Bureau of Internal Revenue (BIR)-compliant ERP system for seamless reporting.
2. **Importance of training:**
   Comprehensive staff training ensured a smooth transition and reduced initial resistance.
3. **Incremental rollout:**
   Testing the system in one section of the warehouse helped identify and resolve issues before full implementation.
4. **Partnership with a local vendor:**
   Collaborating with a vendor familiar with Philippine logistics challenges provided valuable insights and support.

Conclusion

By implementing a WMS, ABC Distributors transformed its warehouse operations, achieving significant efficiency gains, cost savings, and improved customer satisfaction. This case demonstrates the value of investing in technology to stay competitive in the dynamic Philippine FMCG industry.

References

Agyemang, A.A. and Sorooshian, S. (2018). Warehouse management system: A review and research agenda. *International Journal of Supply Chain Management*, 7(4), 260–267.

Akkerman, R. and van Donk, D.P. (2009). Impact of supply chain management on warehouse management: Evidence from the Dutch food industry. *International Journal of Production Economics*, 120(1), 84–95.

Bard, J.F. and Nasiopoulos, P. (2000). A simulation study of warehouse management systems. *Computers & Operations Research*, 27(9), 923–941.

Bortolini, M. *et al.* (2015). A framework for the implementation of a warehouse management system: An application to a case study. *International Journal of Production Research*, 53(3), 927–940.

Kumar, A. and Singh, R. (2016). Optimization of warehouse management system: A case study. *International Journal of Research in Engineering and Technology*, 5(2), 1–8.

Rao, S.S. and Goldsby, T.J. (2009). Warehouse management systems: A framework for decision making. *Journal of Business Logistics*, 30(1), 81–104.

Chapter 7

Automation and Robotics
In Warehousing

Learning Outcome

By the end of this topic, you should be able to do the following:

1. Identify the pallet handling technologies.
2. Recognize the light goods or case-handling technologies.
3. Implement robotics system.
4. Explain the automated guided vehicles (AGVs) and autonomous mobile robots (AMRs).
5. Determine the benefits and challenges of implementing robotics.
6. Explain future trends and innovations in warehouse automation.

1. Introduction

The global trends affecting Asia countries are aging population, labor shortage, urbanization, increasing consumer demand, and complexity of multi-channel fulfillment operations (e.g. e-commerce and hybrid brick and mortar and online channels). These world trends are key factors in increasing storage capacity, fulfillment capabilities, cost, and complexities for warehouse operations for the last 20 years.

Global trends in urbanization and changing consumer demands are affecting the way business operates. Urbanization and industrialization in SEA countries, which have a combined population of more than 675 million, have brought about better household incomes, better education, and an abundance of consumer goods (increase in SKUs) to their societies. Companies need to quickly adapt to the greater consumption of goods and more varieties of goods in fast growing urban cities. Therefore, the warehouses grew larger, to store higher stock volumes and handle larger SKU ranges. This leads to warehouse operations becoming more complex in size and operations to handle faster order fulfillment. Using manual operations of the past, it is increasingly more difficult to keep up with larger order fulfillment requests, or high volumes of orders of small quantities (e.g. e-commerce).

Many business models have also evolved and become successful in SEA in recent decades, such as modern supermarkets or hypermarkets, convenience stores, retail chain stores for health and beauty aids, pharmacies, fashion retailers, furniture, electronics, and white goods. The modern warehouse needs to improve further to handle highly complex operations; such as cold chain management, high-density storage, fast pallet retrievals, mixed case picking with high SKU count, and piece-picking fulfillment. With the shift from physical shopping to e-commerce over the last 15 years, warehouses again needed to transform to handle these very large numbers of small orders and varying size products from a lipstick to a coffee table. Businesses are transforming to compete among themselves and face new business challenges, and retailers are forced to adapt or be left behind.

Warehouse complexities are arising from customers seeking quicker delivery times, greater product availability, and greater traceability from order to delivery. Warehouse automation can enable higher storage volume, high SKU count to be handled, faster velocity to fulfill orders (same-day deliveries), and handle many small orders. Increasingly in Asia countries, many in-house logistics operations and third-party logistics (3PL) companies are turning to warehouse automation to improve their capacities, operation capabilities, and cost-saving measures.

The warehouse automation utilizes one or more hardware technologies, to carry out tasks to fulfill the main warehouse functions of unloading, movement, receiving, putaway, storage, replenishment, picking, packing, staging, and loading of goods and inventory within the warehouse, without or with little human intervention or inputs. Warehouse automation is an integration of different material handling hardware technologies, which are integrated and controlled by Programmable Logic Controller (PLC) systems and operate together with a warehouse control system (WCS) or warehouse management system (WMS). The warehouse staff use the WCS or WMS to interface with the warehouse automation, to execute the goods receiving, product storage, and order picking for shipment orders.

There are many types of warehouse automation systems and technologies (Figure 1). They are broadly categorized as follows:

1. **Pallet handling technologies:** ASRS stacker crane system, pallet shuttle system, pallet conveyor system, and electric monorail system.
2. **Light goods or case handling technologies:** AutoStore system, case shuttle system, miniload crane system, mobile shelf storage system, case sortation system, and light goods conveyor system.
3. **Robotics system:** mixed case picking, carton palletizing, item pick robots, and general purpose, e.g. empty pallet handling.
4. Auto Guided Vehicles (AGVs) and Autonomous Mobile Robots (AMRs).

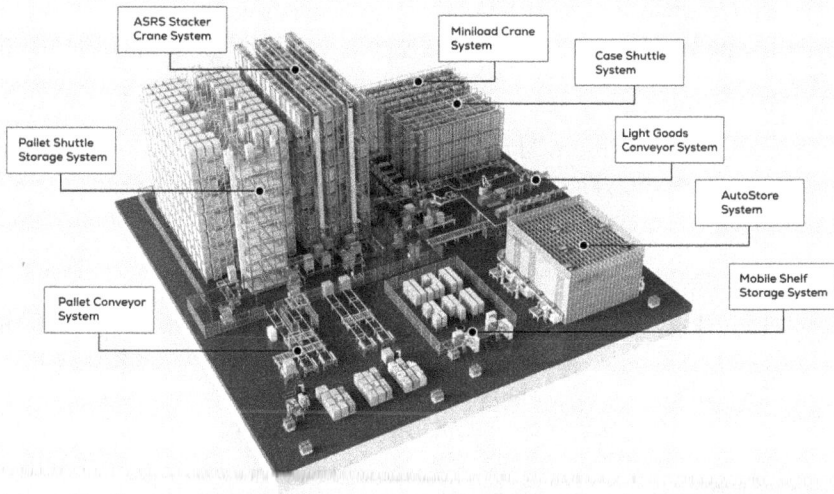

Figure 1.   Technologies available for the warehouse.

Source: Swisslog.

These warehouse automation systems can be viewed in YouTube videos by searching their generic names, to have a better understanding of how the goods and products are handled by automation.

2. Pallet Handling Technologies

1. **ASRS stacker crane system:** The Automated Storage and Retrieval System (ASRS) stacker crane system consists of a crane mast of up to 50m height mounted on a carriage. The Load Handling Device (LHD) with telescopic forks would pick up the product pallets from the conveyor. The stacker crane would lift the LHD as the ASRS carriage runs along the guardrail on the aisle floor. The pallet is lifted and moved to the assigned rack storage position (also up to 50 m height), which the LHD would then put away in the rack position using the telescopic forks. The ASRS racking storage can be single deep, double deep (2 pallets), or multi-deep, depending on the LHD used. The order selectivity enables the ASRS stacker crane system to be a strong choice for high-density storage and can house many SKUs for its flexibility in retrieval. The ASRS stacker crane system can operate in frozen −30°C–50°C temperature. The ASRS stacker crane system can increase storage capacity by >4 times over a manual selective pallet system, for the same warehouse footprint.

2. **Pallet shuttle system (PSS):** The PSS consists mainly of one or more static pallet lifters, aisle shuttles (each with a smaller row shuttle), and a multi-deep racking system. The pallet lifters would bring the product pallet to the assigned racking level and move the pallet to the transfer rack position. The aisle shuttle would pick up the pallet using the row shuttle. The aisle shuttle would move to the assigned row-racking position. The row shuttle would move the pallet out of the aisle shuttle and then move along a C-channel racking to "drop" the pallet at the correct C-channel racking position. The number of pallets can be stored from 5 to 20 pallets typically (subject to design) for a single SKU and single batch, on the C-channel row racking. The PSS has a higher storage density and is capable of higher throughput than the ASRS stacker crane system. However, it is not suitable for a high number of SKUs or smaller batches, such as a retailer's warehouse. The PSS is more suitable for low SKU count and higher quantity per

batch, etc. Plastic resin products and drinks manufacturer, whereby outbound shipments are likely to be single SKUs or batch and in large quantity. The four-way pallet shuttle system is a newer automation, whereby pallet AMRs are used for transport move, put away, and retrieval from pallet racking. The four-way shuttles can move at ground level autonomously with or without pallets and would take the vertical lifter to the upper levels. At the level, the four-way can move along the aisle and then the row locations to drop or retrieve the pallet. The racking system is almost the same as the PSS. The advantage of a four-way shuttle system is that the number of shuttles can be increased according to the throughput.

3. **Pallet conveyor system:** The main function of the pallet conveyor system is to move the pallets from the receiving area to the stacker crane infeed position or PSS lifter position for putaway activities. Vice versa, the pallet conveyor would move the pallet from the crane or lifter outfeed position to the staging area during outbound operations. The pallet conveyor system consists of many different elements which are specifically designed and integrated together to support the required warehouse functions. These elements are roller conveyor, chain conveyor, profile check station, pallet transfer unit, pallet turntable, pallet shuttle car, empty pallet magazine, pallet exchanger, pallet load stacker, low pallet lifter and vertical pallet lifter. The pallet conveyor system with different elements can be configured as a pallet sequencer, for outbound shipments to be arranged in a sequence for First In Last Out (FILO) loading of containers.

4. **Electric monorail system (EMS):** The EMS consists of individual-driven vehicles mounted or hung on an overhead monorail system. These EMS vehicles are connected to a power bus bar and communication bus bar fixed on the overhead track. The locations are also determined by serialized barcodes pasted on the overhead rail track. The EMS is used to transport pallets across long distances and has high movement rates depending on track length, number of EMS vehicles, number of pickup stations, and drop-off stations. The EMS can work together with pallet lifters for goods pallets to move to different warehouse levels. The EMS can be used to transport goods from one building (factory) to another building (warehouse) connected by an overhead bridge.

3. Light Goods or Case Handling Technologies

1. **Auto store system:** AutoStore system is a cubic storage system without any operating aisles inside the system. AutoStore system consists of robots, storage bins, a grid system, operating ports, and an AutoStore control system. The robots would "dig and pick" the required product bins to the ports for the picker to pick the items. Similarly, the robots would present the empty bins to the putaway ports for operators to put away products into the storage bins. The storage bins are stacked on top of each other, typically up to 16–24 stacks, depending on bin height (330mm or 220mm, respectively). The maximum load weight for each bin is 30 kgs. The advantage of the AutoStore system is it supports carton picking and inner or piece picking. The picking performance is high depending on the number of robots and operating ports. The picker is stationary without having to walk through the whole warehouse to pick up the items. Due to its cubic storage concept, the space occupied by AutoStore can be about 25% of the area occupied by rack shelving system or conventional racking system for storage (Figure 2).

2. **Case shuttle system:** The case shuttle system is the smaller version of the pallet shuttle system and consists of case shuttles, case vertical lifters, racking systems, case conveyors, and control systems. Cartons or tote boxes, of up to 50 kgs, are moved into storage in the racking system by case lifters (vertical movement) and by the case shuttle (horizontal movement) to the case storage locations in the racks. The storage height can be up to 24m in height, giving a higher storage density than

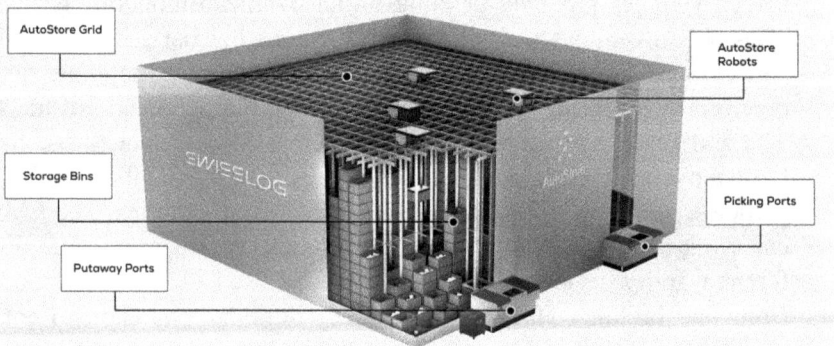

Figure 2.   Autostore architecture.

Source: Swisslog.

conventional shelving storage of 12m height. The cartons or totes can be stored from 2 deep to 6 deep, depending on the sizes. The case shuttle system is a high-density case storage system and has high performance picking through integrated picking station designs. The case shuttle system can also be used as a shipment consolidation buffer, after the products and orders are picked and packed into shipment cartons.

3. **Miniload crane system:** The miniload crane system is the smaller version of the ASRS stacker crane system and similarly consists of the crane mast, loading handing device (case or tote), case storage racking system, case conveyor system, and control system. The cases are put away or retrieved using the crane mast and LHD to and from the case storage locations in the racks. The storage height can be up to 24m in height, higher density than conventional shelving storage. The cartons or totes can be stored from 2 deep to 4 deep, depending on the sizes. The throughput performance of the miniload crane system is about 2–3 times lower than the case shuttle system and similarly uses integrated picking station designs. The main advantage is its ability to handle loads up to 250 kgs and is more flexible in larger size loads (Figure 3).

4. **Mobile shelf storage solution:** A mobile shelf storage system is a storage and goods-to-person order fulfillment system consisting of portable standalone shelves, autonomous mobile robots (AMRs), picking stations, and fleet manager systems. This system is primarily designed for multi-channel e-commerce fulfillment and retail operations. A fleet of AMRs delivers the mobile storage racks to workstations for case or item picking and putaway. The main advantages are

Comparison	AutoStore	Mobile Shelf Storage	Case Shuttle System	Miniload Crane
Storage Density	Very Good	Average	Good	Good
Throughput	Good	Good	Very Good	Average
Scalability	Very Good	Very Good	Good	Average
Redundancy	Very Good	Very Good	Good	Average
System Height	Good	Average	Very Good	Very Good
Weight & Size	Average	Very Good	Good	Very Good

Figure 3. Comparison of Light Goods Handling Technologies
Source: Swisslog.

flexibility in carton size or elongated shape (e.g. broom or ironing board), simple design and operations, and scalability in expansion to cater to business growth. The mobile shelf storage system has a lower storage capacity (usually up to 2m height) and the picking performance is comparable to the AutoStore system and case shuttle system. However, it has lower capital investment costs.

5. **Case sortation system:** Sortation is the process of identifying items on a specially designed conveyor system, after induction (putting cases on the sortation system) and diverting them to a specific destination. Automated sorters leverage on barcode scanners, sensors and divertors, to sort the cartons automatically with no human intervention. Sortation systems are used in three main functions within the warehouse: (1) batch orders case picking to sort to orders, (2) flow-through of products from receiving to orders, and (3) many parcels or shipment cartons to sort to destinations or shipment routes. There are many types of sortation systems: pusher sorter, pivoting arm sorter, pop-up wheel sorter, sliding shoe sorter, push tray sorter, tilting tray sorter, and cross-belt sorter. The type of sortation system selected would depend on the sorting performance required, the type of products (packaging, size, and weight), and the number of sorting lanes.

6. **Light goods (or case) conveyor system:** The main function of the light goods conveyor system is to move the totes, crates, or totes from the receiving area to the storage in AutoStore, case shuttle system, or miniload for putaway activities. Vice versa, the light goods conveyor would move the totes, crates, or totes from the storage system to the order fulfillment operations areas, which could be the picking station, case sortation system, packing station, or palletizing station.

4. Robotics System

Robotics system: Robotics system generally refers to articulated arm robots, which are similar to human arm movements. They generally have 5 or 6 axes of movements which gives them flexibility through the various points in the three-dimensional space. An end effector or "gripper" is attached to the robot arm, which is designed to perform one or more specific tasks, such as gripping, suction, picking, mechanical clamping, and holding. General uses of robotics in warehouse operations are loading or stacking of empty pallets, carton depalletizing or palletizing, case picking, item or piece picking, and sorting (Figure 4).

Figure 4. A robot arm for warehouse operations.
Source: Swisslog.

One application of robotics is fully automated order picking of mixed case pallets. This advanced solution allows store-friendly pallets to be automatically picked in a "lights out warehouse" and be built up without human labor. The case buffer system would store the cases in case racking, and the case shuttle system or miniload would bring out the required cases. These cases are then sequenced on the light goods conveyor systems according to the palletizing sequence software. After that, the cases are then palletized by the robots with grippers, onto the pallet in a sequence of heavy to lighter goods and large cartons to smaller cartons.

The item pick robot is a smaller robot picking system, which is ideal for e-commerce, retail chains, pharmaceutical retail, and medical logistics for small item picking of less than 5 kgs. The item pick robot gripper is usually designed with three small fingers picking or with small suction pads to handle small items. The robot is designed to handle high repeated piece picking and error-free fulfillment. The robot can pick and place products continuously, from a source tote into an order bin or carton, working together with the AutoStore system or case shuttle system or miniload system. This allows companies to better cover seasonal peaks, such as festive holidays and events, thus reducing operating costs in the long term and overcoming staff shortages.

5. Automated Guided Vehicles (AGVs) and Autonomous Mobile Robots (AMRs)

In the past, AGV referred to a class of automated guided vehicles which operated in a "fixed routing" and "tasking" environment, while AMRs move a lot more autonomously and can change their routes and taskings dynamically in real time. In present-day warehousing, the fleet managers and navigation of AGVs and AMRs are so much enhanced that the two terms are used interchangeably.

The AGVs and AMRs are further categorized by design and function: (1) fork AGVs which are similar in design to forklift and reach trucks, (2) platform AMRs which have flat "oyster" shape and carry their load by lifting or lowering, (3) tug AGVs which pull their load mounted on wheels (e.g. trolleys), and (4) conveyor AMRs with roller or chain conveyors on them. The fork AGVs (forklift, pallet truck, or reach truck) can operate in three dimensions by putting the pallets on the floor and storage racking, while the rest are generally used for transport movement moving the loads (pallet or case) from point to point.

The AGVs and AMRs are flexible warehouse automation, as they are not fixed installations unlike other warehouse automation. Their implementation does not "restrict" movement, such as conveyors and racking systems which are permanent fixtures occupying warehouse space. The costs of AGVs and AMRs with the fleet manager are generally lower than other capital investments in other fixed warehouse automation. The number of AGVs or AMRs can be increased accordingly to business growth and increased throughput. AMRs can be used in the mobile shelf storage system moving the racks, replacing case conveyors between the picking stations and the case shuttle system or miniload crane system, and in the sortation process of cases or order totes.

6. Benefits of Warehouse Automation

The benefits of warehouse automation can be broadly categorized into the four Cs and D as follows:

1. **Capacity:** to handle more storage volume.
2. **Capabilities:** to do more, faster, and better.
3. **Costs:** reduction in warehouse space used and improved labor productivity.

4. **Complexities:** handle complex warehouse operations, for example, e-commerce and omnichannel.
5. **Digitalization:** system-generated data enable descriptive visualization, optimization, prediction, and prescription of warehouse operations.

Capacity:

The storage volume of warehouses in Asia countries has grown tremendously due to population growth, urbanization (bigger and denser cities), income growth, greater purchasing power, and growth of many consumer goods and higher consumption. There is growth in retail business or modern trade in grocery supermarkets and hypermarkets, large furniture retail boxes, convenience stores, pharmacy stores, health and beauty aids (HABA) chain stores, and e-commerce or online purchases. Goods and products are moving in greater volume and faster rate into the warehouses, picked to orders, and then distributed to retailers or consumers.

The ASRS stacker crane system utilizes 50 m racking height for the storage of pallets, compared to 12 m height for a conventional selective pallet racking system. The ASRS stacker crane aisle width is 1.65 m, instead of 3.2 m for the selective pallet racking system. The ASRS stacker crane system typically stores pallets in double-deep racking instead of single deep, saving another 3.2 m width aisle space. For these reasons discussed, the ASRS stacker crane system utilized about 25% or one-quarter of the warehouse space used by the selective pallet racking system and reached truck operations for the same quantity of pallets in storage.

The AutoStore system provides cubic storage for cases, inners, or pieces in high storage density but without the operating aisles of the rack shelving system or pallet racking system. The AutoStore system utilized about 25–30% of the warehouse space of the conventional racking system. This can translate to savings in warehouse rental and higher savings in aircon warehouse space rental.

Both the case shuttle system and miniload crane system can be designed up to 24 m height, which is double the height of the conventional rack shelving system of 12 m. Therefore, warehouse automation enhances the storage capacity of a given warehouse land area. In support of a manufacturing plant scenario, the increased warehouse storage capacity enabled the factory to produce more without having to shift the factory to a bigger new site or to re-locate the finished goods to another offsite warehouse which involves double handling of products in transport.

Capabilities:

The capabilities or the high performance of the warehouse operations refers to the ability to do more within a given warehouse space and the ability to do faster for shorter order fulfillment timings and service quality in order accuracy. Generally, the ability to do more and the ability to do faster are related. However, in some automation designs, they can be mutually exclusive. Examples are stacking two or more lines of conveyors (pallet or case) on top of each other, giving a higher movement rate using the same warehouse space. In many automation designs, mezzanine levels are used on top of the staging area for picking stations. The cartons or orders are transferred immediately downward to the staging area for dispatch, whereby it is not possible or too difficult to design in manual operations.

In pallet handling operations, pallets can be moved, put away, and retrieved at a higher rate than forklift or reach truck operations. Pallets can be moved around the warehouse using AMRs or AGVs, which numbers can be scaled to meet operational needs. The overall performance (number of pallets moved) is higher due to its fleet manager software, matching the pallet drop-off with pickups and safety sensors allowing more AMRs to be deployed in the same area. The putaway and retrieval by ASRS stacker cranes are generally two times the performance of a reach truck operator, while the PSS can be scaled to perform greater than 150 pallets in and out per hour, depending on the automation design.

In the case of inner or piece picking operations, AutoStore, case shuttle system, miniload, and mobile shelf storage system can enable a warehouse staff to pick more than 180 cartons or pieces per hour, versus a typical rate of 80–120 cartons or pieces per hour for manual picking in pallet racking system. Automated picking systems increase labor productivity by more than two times.

Cost:

The annual cost of automated warehouse operations can be lower than manual operations, mainly by reducing the warehouse space used, lowering the number of warehouse staff, and lowering the number of forklifts or reach trucks. In the earlier discussion on the ASRS stacker crane system, it utilized about 25% or one-quarter of the warehouse space used by the selective pallet racking system and reached truck operations for the same quantity of pallets in storage. This 75% warehouse space saving is translated into warehouse floor construction and therefore the cost of building construction. The 75% cost saving in new warehouse construction would contribute to the ROI payback period calculations. The capital investment spent on warehouse construction can be greater than the amount spent on the automation investment for a large warehouse. The higher capital investment in building construction also attracts a higher interest for bank load. This saving can also be in the form of monthly warehouse rentals in the case of sale and lease back arrangements or public warehouse rentals.

The AutoStore system utilized about 25–30% of the warehouse space of the conventional racking system. This can translate to saving in warehouse rental and even higher savings in aircon warehouse space rentals. For aircon warehouses, there is additional saving in electricity used due to the smaller footprint occupied by the AutoStore system. The cost saving from labor productivity is significant, as labor productivity can be increased from 1.8x to 3.0x higher for picking activities. In the ASRS stacker crane system, there are significant savings from operators and reach truck equipment for putaway, pallet replenishment, and outbound for full pallet picking activities. The ASRS stacker cranes can last 2.5x–3.0x longer than the reach truck equipment also (about 5–6 years depending on usage). The labor savings from automated warehouse operations increased by many times more; for two shifts of warehouse operations where two sets of warehouse staff are needed and for seven days per week warehouse operations where additional manpower is recruited to cover the rest days of some workers.

For frozen or chilled warehouses, the 25% footprint of the ASRS stacker crane system translated to smaller air volume and smaller surface area of the walls and roof. Therefore, other than construction savings from a smaller warehouse footprint, there are more savings from the refrigeration capital equipment. The small air volume means the refrigeration capacity and its capital investment can be smaller. There is a lesser running cost of utility every day for a smaller air volume to cool the warehouse. The smaller surface area of walls and roof of automated frozen or chilled warehouses means less capital investment in insulation panels and installation. There is lesser heat load from direct sunlight or indirect light from surrounding the walls and roof. This would reduce capital investment in the refrigeration system and also the running cost of utility.

Complexities:
As mentioned in the introduction, many business models have also evolved and become successful, such as modern supermarkets or

hypermarkets, convenience stores, retail chain stores for health and beauty aids (HABA), pharmacies, fashion retailers, furniture, electronics, and white goods. These modern business models have highly complex warehouse operations, high SKU or product count, high-volume mixed case picking, inner or piece picking fulfillment, high-density storage, fast pallet retrievals, a very high number of small orders (e.g. e-commerce), and cold chain management.

Warehouse automation designs can integrate various functions of the warehouse together; for example, a pallet handling system for storage and full pallet picking, a case handling system to handle case picking, and another system for inner or piece picking, according to the product characteristics and order quantities analysis. The different picking operations (pallet, case, and inner or piece) of a single order can be consolidated by warehouse automation, just before the dispatch or shipment time. For e-commerce fulfillment centers, warehouse automation can receive, store, put away, pick, and pack for more than 1 million SKUs and yet does not need the enormous warehouse space.

For retail chains of many small stores, such as convenience stores, pharmacy stores, HABA stores, and personal electronics stores, the light goods handling technologies can consolidate the order totes into a case shuttle system or miniload system up to 24m height. This reduces the large area needed for staging these orders and the time needed to locate and consolidate each tote or pallet. These systems function as a "Vertical Sortation System" or "Vertical Staging Area". The order totes of specific store orders and of a shipment load (or a truckload) are then called out to the loading bay about 30 mins before truck arrival, thus simplifying the consolidation process and shipment loading.

For the manual frozen warehouses, the reach truck operators and case pickers would need to work in very cold temperatures lower than $-18°C$ for many hours. The warehouse automation for frozen warehouse can place product pallets in $<-18°C$ ASRS stacker crane system and then automatically replenish the $<-18°C$ case buffer storage racking using a case shuttle system or miniload crane system. The required frozen product cases are then brought out to a 5°C chilled area for palletizing by workers or robots. The order pallet is then bought back to $<-18°C$ ASRS for staging or consolidation. This would simplify the workflow for the frozen warehouse staff, such that they do not need to work in the $<-18°C$ environment.

Digitalization:

A significant benefit of warehouse automation, which contributes to advancement of Industry 4.0 implementation, is digitalization of warehouse operations. Warehouse automation digitalizes the material flow and the workflow within the warehouse, and the capturing of many important data of the warehouse operations automatically. Many WMS and WCS have 3-D Visualizer of the warehouse layout and automation in "live' mode, whereby movements of materials (pallet or case) & automation equipment (e.g. ASRS stacker crane or AMRs) are also animatedly displayed on screen (Figure 5).

The 3-D visualizer or the "Digital Shadow" described the warehouse operations in real time to the warehouse management staff. Warehouse management staff do not have to walk physically to "tour" the entire warehouse to gain understanding of the warehouse operations. The warehouse supervisor now works with two computer screens; for example, one screen is for processing an order on the WMS and releasing the order and the other screen shows animatedly the 3-D pallets coming out of the racking storage onto the conveyor and forming at the staging area. He can then check on the WMS module or the visualizer whether the order of the required pallets is completed.

Figure 5. A 3-D Visualizer for real-time monitoring for operations.

Source: Swisslog.

If there were events such as stoppages or equipment issues, that portion would be highlighted "red" and a message would inform the supervisor of the status. The supervisor can liaise with the technical staff via voice communications on RF terminals or walkie-talkies and instruct where to check the possible fault. The data collected by the WMS or WCS can be reviewed or the animation video be played back for operations review and optimization of the warehouse operations. Business Intelligence (BI) can also be interfaced with the WMS or WCS to enhance the review of the past data, for prediction of increased throughput volume during festive seasons.

6.1 *Challenges in implementing warehouse automation*

1. **High capital investment and return on investment (ROI):** As it comes to technology adoption, the main barrier to companies in Asia is typically the high initial investment cost. But many companies are starting to realize that there is a business case to upgrade their warehouse design. The cost savings in building infrastructure and higher manual operating costs can pay off the investment for warehouse automation. Companies can form an automation project team, which comprises staff with different skill sets, operation managers, industrial or automation engineers, and software system analysts. The automation project team could review the current warehouse operations, conduct the SKUs analysis and order management analysis, and receive further training and education in warehouse automation. They can be supported by a finance staff to explore the areas of improvement and the possible cost savings by transforming the current warehouse operations with automation. Simple and small automation systems typically have a payback period of 3–5 years, while more complex and larger automation systems would expect a longer payback period of 7–12 years. With good understanding of the ROI and payback period business case, company management is more acceptable of the implementation of warehouse automation.

2. **Selecting the right solution and design:** There is yet to be wide adoption of automation in warehouses in many Asia countries. The companies do lack strong understanding of the various automation systems and to select the right automation that can best support their warehouse operations. The automation project team could attend the

warehouse automaton training or conferences and conduct discussions with system integrators or suppliers to gain better knowledge. The automation project team can also visit the warehouse automation used in the same type of distribution centers and the same products in the same industry, e.g. grocery retailer, HABA retailer, drinks manufacturer, or a semi-con distribution center. The automation project team can also engage consultancy companies who have a good track record of consulting and implementing warehouse automation. This would provide further confidence level that the right automation solution is implemented.

3. **Logistics staff expertise and education:** The automation project team would need to work on change management plan and staff education program to adequately prepare warehouse operators for the big change from manual warehouse operations to automation-driven operations. The project team can provide a series of training sessions on warehouse automation from a basic understanding of what automation would do to how the warehouse workers would work in the new automated operations environment. At this stage, it is also useful to bring a bigger group of warehouse staff to visit a similar automation system at another site. Sharing sessions can allow the staff to interact and discuss with the other warehouse staff, who have gone through the implementation of warehouse automation in their operations.

4. **Training the automation-operator interface:** The company and management staff would need to prepare and educate all warehouse staff, to be ready for the new warehouse automation operations. Many aspects of the warehouse processes and activities are different from manual operations. In change management, it is important for the warehouse staff to understand why there is a need to implement a new automation system and how automation benefits the workers. The workers, especially forklift operators, are often apprehensive about potential job loss and would need re-skilling to adapt to the new roles and working environment. Regular training to introduce automation systems and processes to the warehouse staff is important. The use of animation videos or actual system operation videos would assist them in understanding their new job functions better. The more adaptable workers to new technologies and software can be given buddy leader roles to assist other weaker workers. The company can also involve more staff (e.g. supervisors) to go on visits to similar automated warehouse operations and interact with workers there.

5. **Engineering and technical staff:** The modern warehouse organization would include an automation team, comprising an automation manager or engineer with vocationally qualified technicians. The automation team is to support and maintain the automation system and provide the on-site fast response to any faults or system "hang" issues. The team would also liaise with the system supplier for periodic preventive maintenance or enhancement and upgrades. The automation manager/engineer is an important member of the warehouse management team, advising on system capabilities and performance, and training warehouse operators to work with the warehouse automation.

6. **Transition and preparation:** The automation project team, after concluding the design of the automated warehouse and signing off the contract, would need to have user specification workshops with the automation supplier on software and detailed engineering design and layout. During these user specification workshops, the project team should take note of the differences in processes, data and information, warehouse equipment, and handling by different operators. From this understanding, the project team must prepare a list of preparation tasks to prepare the manual warehouse operations to transition to the automated warehouse operations. Many of these preparation tasks involve standardization of workflows and materials. One example is the printed pallet label with the standard format and barcodes. The pasting of this label shall also be standardized to a specified position (height above ground, side and corner/middle) on the pallet for reading by fixed barcode scanners on conveyors. The wooden pallet or plastic pallet which would travel on the roller or chain conveyors would need to be of a standardized design and load weight. Many of these preparation activities can be carried out during the manual warehouse operations while awaiting the warehouse automation system to "Go Live" during implementation.

To help companies make the most out of their automation investment, it is essential to select an automation partner that has extensive global or regional experience, with highly knowledgeable personnel and a willingness to enter into a long-term partnership to the mutual benefit of both parties. Automation suppliers can partner with and support the company for every step of the design, installation, preparation for Go Live, after-sales support, and upgrade journey. It shares its knowledge and showcases the best practices from its successful global or regional project implementations.

The automation partner should have a strong presence and customer support team regionally or in the country to support the automation project and operations. It is preferred that the engineering and design staff are localized (or regional) and have a strong track record to demonstrate their experience and commitment to the local market. Their localized expertise can help tailor solutions that are optimized for specific local needs, ensuring maximum operational efficiency for warehouse operations. Their established network and knowledge of local regulations and infrastructure can facilitate smoother operations and better adaptability to local situations.

The customer support staff should be in the country to respond quickly to system issues and expert troubleshooting for complex problems. The localized support team can access higher engineering expertise (global or regional team) for complex issues. An automation partner with a highly skilled engineering team ensures effective problem-solving and minimizes downtime.

The automation partner should have a good track record of implementing several warehouse automation projects in the country or neighboring countries. These would be good reference sites for the automation project team to visit and gain knowledge from other automated warehouse operations. The automation partner should have good WMS (WMS partner or in-house WMS) and software intelligence to integrate and manage the automated operations and capabilities. Again, the automation project team can visit the reference site and interact with the actual automation warehouse users. The automation partner should have cutting-edge automation technologies in various sub-systems; robotics, AMRs, high-performance picking stations, and inner and piece-picking capabilities so that the automated warehouse can be further enhanced or upgraded.

7. Future Trends and Innovation in Warehouse Automation

Warehouse automation is continually evolving to be better and faster. New innovations are elevating warehouse automation to higher levels of technologies. Flying drones are used in inventory stock take or cycle count, during the night hours or non-operating hours. The flight paths of the flying drones are mapped by the fleet manager to "scan" the stock take or cycle count locations in its flight path. The vision camera of the drone captures the inventory barcodes on the pallets and links them to rack location barcodes. The data captured is fed back in real time to the WMS, which then matches the results with the WMS Inventory. This would typically take about 1–2 warehouse staff to execute, instead of deploying many staff and equipment for many hours of stock take. In the future, flying drones can be used to carry average-size cartons or products to the consolidation directly or for the execution of urgent orders but not foreseeable for mass execution which requires many drones. The safety of flying drones is still a major concern when there is drone failure.

The R&D or proof of concept deployments of humanoid robots or mobile arm robots in many laboratories would eventually enable these state-of-the-art innovations to be operational in the actual warehouse environment. The warehouses would become more intelligent and efficient and adapt to changing demands. The arm robotics are mounted on AMRs and are capable of doing piece picking of up to 5 kgs, or carton picking of about 15 kgs. Many types of piece-picking mobile arm robots are starting in a warehouse, as these arm robots are smaller and the AMR supporting it is also smaller, the order tote is also housed on the same AMR.

As the load size and weight reach a carton of about 15 kgs, the arm robot becomes bigger and heavier. Therefore, the AMR has to be a heavier platform to counter the force of a swinging arm robot and also provide greater battery power. Another AMR carrying the order pallet would receive the picked cartons from the first mobile arm robotics. The early application of Artificial Intelligence (AI) in warehouses is robotics picking. The AI-based robotics arm not only is able to identify and pick a wide variety of items, but it also has machine learning to optimize its gripping points to pick unknown items. Then, the robotics arm will be able to recognize new or unknown items and decide on the optimal approach to grip the item.

The most difficult challenge in warehouse operations is automating the stuffing (loading of loose cartons) and unstuffing of sea containers. The confined space (8 ft. width by 8 ft. height — external dimensions) and varying carton sizes limit many automation options. The deployment of humanoids and the mobile robotic arm system are making stuffing and unstuffing operations possible in the future, to work in a confined space of sea containers. The use of AI would also enable the loading or unloading of varying carton sizes in the right sequence or segregating the same SKU cartons onto the same pallet, as humanoids and mobile robotic arm systems can work in more flexible actions and methods.

The concept of self-learning smart warehouses is introduced, in which an AI-managed warehouse driven by data and robotics technology would be able to adjust and optimize warehouse processes, based on changes in customer demands. The AI would integrate with WMS and Warehouse Equipment System (WES) to derive insights from a data-rich warehouse environment. Forecasting of inventory, "hot spot" analysis of operational flow, slotting of picking, and AMR taskings can be employed to ensure optimal space usage and streamlined operations. Big data, enabled by the digitalization of warehouse operations, offers valuable insights into performance, bottlenecks, and operational trends within the warehouse. Data analytics can be implemented to monitor key warehouse performances (e.g. pick rate and dock change-over duration), identify inefficiencies, and make informed and data-driven decisions to implement continuous improvements to the warehouse operations.

AI also helps improve inventory management of an automated warehouse, automated warehouse planning and even forecasting can be achieved. By analyzing the historical transactional and operational data, an AI-integrated WMS would be able to automatically initiate various warehouse operations, such as order releasing and picking scheduling according to the current workload and time periods in the warehouse. With integration into external factors such as marketing campaigns or weather conditions, it is even possible to make predictions about customer ordering behavior and allow further optimization of warehouse operations. The future warehouse is intelligent enough to be able to adjust its capacity and capability to accommodate the changes in market demands throughout the year.

To fully maximize the investment in warehouse automation, all hardware and software must perform at optimal conditions. Predictive maintenance utilizes various IoT sensors (e.g. heat, vibration, noise, and temperature sensors) to capture the real-time status of measured variables. Combined with historical data, predictive maintenance plays a vital role in this by proactively identifying potential failures before they occur. Leveraging data analytics, machine learning, and specialized sensors predictive maintenance solutions continuously monitor equipment performance metrics and issue alerts for any deviations. AI would help detect potential failures of mechanical parts in advance, thus minimizing the risk of unplanned downtime. This enables timely repairs, ensuring uninterrupted operations, enhanced efficiency, and improved safety.

There are many more potentials for AI making its way into warehouse automation. For example, an AI chatbot can be fed with relevant knowledge to serve as an automated assistant for troubleshooting. Integrating with "Digital Shadow" mentioned earlier, an AI chatbot would be able to provide first-level support to allow faster recovery. The chatbot could also capture relevant data automatically for further root cause analysis.

8. Case Study: Logistics Company G

Logistics Company G is a Southeast Asia logistics subsidiary of a grocery supermarket retailer. In the late 1990s, G implemented the WMS using RF technology to improve the efficiency in handling dry goods and grocery items for supermarket replenishment daily. The WMS and RF system improved faster system-driven operations for its grocery warehouse, with higher outbound throughput. As the grocery retailer's SKU range grew by over 5,000 products in the next six years, Company G saw the need for further automation to upgrade its warehouse on its same industrial land.

In 2007, Company G embarked on its automation journey by implementing an automated case sortation system. This system, combined with the WMS, enabled batch and zonal picking, a pick-to-light system, and automated sorting, making the warehouse more efficient and adaptable to growing supermarket orders. The grocery cartons are dropped off at conveyor infeed stations after picking, and finally, the automated sortation system would sort the grocery cartons or cases to supermarket orders into their respective sorting lanes. The cartons or cases are then palletized by workers to be delivered to the supermarket.

By 2011, with strong retail sales expansion and more product variety, G formed a project team to conduct a five-month project study and also visited several grocery retailers' automated distribution centers in Europe. Later, Company G invested in a highly advanced distribution center, introducing more than 50,000 pallet Automated Storage and Retrieval Systems (ASRSs) with CaddyPick system technology. This advanced distribution center features 20 stacker cranes of 42m height and an electric monorail system allowing for rapid pallet movement of more than 400 pallets per hour. The automated CaddyPick system is integrated into the ASRS module at the mid-level racking system. The integrated ASRS stacker crane system has three levels of 11 picking aisles each, which significantly increases the throughput to 120,000 cartons daily, boosting picker productivity by over 20%. This new distribution center also deployed six industrial robots to handle empty pallet stackings and loading of order pallets for picking.

In 2017, recognizing the rise in e-commerce, G operationalized the standard AutoStore solution to automate order picking for e-commerce fulfillment. The e-commerce AutoStore improved the storage density and increased online order handling by more than three times. Then, in 2018, the e-commerce fulfillment center was further enhanced, expanding the same AutoStore system at the same warehouse site. The enhancement upgrade was done while the daily operations were ongoing. The enhanced e-commerce fulfillment center uses an expanded AutoStore system, adding three times more robots. The e-commerce fulfillment center also integrated a case shuttle system for vertical "staging" or "buffering" of e-grocery order totes. The robots palletized the order totes and the AGVs would move the order pallets with totes, to the staging area for loading into trucks. This stepwise approach allowed Company G to steadily grow its automation capabilities and human talents, transforming into fully optimized and scalable logistics operations.

9. Case Study: Douglas Pharmaceuticals using ASRS from Kardex

Amid rising real estate prices, energy costs, labor shortages, and ongoing economic uncertainty, automation is becoming a critical tool for companies seeking to optimize operations. Modern warehouse automation systems are offering key advantages, such as improved space utilization, scalability, efficiency, and sustainability—qualities that are essential to remain competitive in today's business landscape.

One class of solutions gaining traction is the cube-based storage and retrieval system. These systems are designed to maximize storage density by stacking containers vertically within a compact grid. Robots traverse the grid, retrieving and delivering items to workstations. This approach allows companies to scale their warehouse operations with reduced dependency on manual labor and without the need for major expansions or high capital expenditure.

These systems are often implemented in partnership with automation providers who support businesses in assessing their current needs, running simulations, and developing tailored implementation strategies. In many cases, such providers also offer software and complementary tools to optimize automation performance and ensure ongoing adaptability as requirements evolve.

As organizations across the Asia-Pacific region continue to face mounting pressures, particularly in warehousing and logistics, there is a growing demand for solutions that are cost-effective, space-efficient, and flexible. Rising costs of construction and industrial property rentals are particularly significant challenges. For example, data from sources such as Interest.co.nz and Statistics from New Zealand show sharp increases in warehouse construction costs over the past five years, placing financial pressure on businesses planning expansion.

In New Zealand, industrial property markets are especially tight. According to reports from real estate analysts, cities such as Auckland, Wellington, and Christchurch have seen notable hikes in industrial net prime rents during 2023 and 2024, driven largely by a demand–supply imbalance.

Singapore has experienced similar trends. Market reports indicate a 10.5% year-on-year increase in industrial rents in 2024, following a 7.4% rise in 2023. These increases are forcing companies to re-evaluate how they manage space and resources.

Meanwhile, in the Philippines, labor and skills shortages are among the most pressing operational barriers. According to the World Economic Forum's *Future of Jobs 2025* report, 67% of executives in the country cite the skills gap as the top challenge to business transformation, surpassing the global average of 63%.

Singapore, too, is focusing on workforce development. The same WEF report emphasizes the importance of upskilling and the adoption of process automation as strategies for future-proofing business models in an evolving market.

Overall, as labor constraints and operational costs continue to rise across the region, automation is emerging as a vital strategy. It enables businesses to improve productivity, reduce reliance on manual processes, and develop scalable systems that support long-term growth. By leveraging technology, companies are better positioned to manage the economic and logistical pressures reshaping the warehousing and logistics sectors.

9.1 *Douglas pharmaceuticals sees 68% throughput gain and 30% more storage with warehouse automation*

In 2024, Douglas Pharmaceuticals—one of New Zealand's largest privately owned pharmaceutical companies—implemented the country's first cube-based automated storage and retrieval system at its West Auckland facility. The project addressed key logistical challenges, resulting in a 68% increase in throughput and a 30% expansion in storage capacity, all while preserving operational efficiency within a limited warehouse footprint.

9.1.1 *The challenge*

Founded in 1967, Douglas has become a respected name in the global pharmaceutical industry. As well as producing prescription drugs, Douglas develops, manufactures, and distributes a wide range of consumer health and wellness products. As a growing company with both domestic and international customers, it faced the common issue of its manual, highly utilized warehousing operations becoming a bottleneck, particularly for its fast-moving consumer healthcare range.

To increase its storage capacity, Douglas considered a range of options:

- **Expanding the current warehouse:** This would involve reconfiguring the road network at the headquarters site, but the logistical complexity and high costs made this option less attractive.
- **Purchasing a new storage and fulfillment facility:** This would incur the expense of acquiring new property and setting up another facility, which would be costly and geographically challenging.
- **Holding product off-site:** Storing product away from the main campus posed logistical difficulties and increased operational complexity.

These options were weighed against the backdrop of Douglas' desire to remain on its current West Auckland site, which also houses its warehouse, distribution, sales, marketing, and corporate head office. The company is deeply rooted in the local community and expanding physical space was not a feasible solution due to prohibitive costs and low returns.

In light of these challenges, Douglas looked toward automation as a potential solution. Automation would provide a way to avoid the scaling issues inherent in manual operations while giving the company the capacity needed for future growth. The manual configuration had become inefficient, with employees covering large distances and "getting in one another's way." Furthermore, automation promised a substantial increase in throughput and operational efficiency, making it an attractive solution for Douglas' needs.

9.2 *The solution for Douglas*

After evaluating the various options, Douglas turned to Kardex, a global partner for AutoStore, to implement a high-density storage solution that would meet its growing capacity needs. The success of the project was largely attributed to Douglas' detailed understanding of its operations, which played a crucial role in ensuring the effective implementation of automation.

Kardex collaborated closely with Douglas to gather and analyze data, running simulations to demonstrate how the AutoStore ASRS (Automated Storage and Retrieval System) could solve its logistical challenges (Figure 6). Through meticulous planning and modeling, they were able to

Figure 6.   ASRS for the company.

identify a solution that would provide increased efficiency and scalability while remaining cost-effective. Additionally, the teams worked on change management, developing internal champions to facilitate a smooth transition for employees.

No significant modifications were required to the existing warehouse or its power infrastructure. To prepare for the installation of the AutoStore grid, some temporary office space was cleared, and the warehouse floor was ground down by CoGri (New Zealand), following testing by Face Consultants (New Zealand). This process was carried out while the existing manual operation continued, minimizing disruptions. The AutoStore system at Douglas' Auckland facility is a high-performance solution featuring 15,000 high-density bins, managed by 13 high-speed robots, 3 CarouselPorts, and 1 ConveyorPort.

Although Kardex offers its own proprietary Warehouse Management System (WMS), in this case, the team integrated Douglas' existing SAP system to create a holistic warehouse management solution. This integration included an integrated front-end for warehouse operators, designed to maintain a similar look and feel to the previous picking system, ensuring a seamless transition and ease of use.

9.3 *Effective collaboration yields success*

The successful outcome of this project reflects effective collaboration among all stakeholders involved. Completed on time and within budget, the implementation met all key objectives. The achievement was the result of careful planning, close coordination, and the efficient use of automation technology. Jason emphasized the strategic benefits for Douglas Pharmaceuticals as the first company in New Zealand to adopt this type of system, noting, "It's a valuable marketing tool for Douglas. They now have a significant logistical advantage that sets them apart."

The automated storage and retrieval system implemented at Douglas Pharmaceuticals has delivered significant improvements in warehouse operations, particularly in storage density and processing efficiency. Picking times have improved by up to four times, overall throughput from order receipt to dispatch has increased by 68%, and storage capacity has expanded by 30%. The compact grid system currently occupies just 10% of the total warehouse space, allowing ample room for future expansion. Designed for scalability, the system is expected to support at least five years of projected growth, with the flexibility to add more robots and workstations as demand increases (Figure 7).

Figure 7.   Topview of the ASRS from AutoStore.

Through detailed virtual simulations, Douglas Pharmaceuticals was able to visualize how the new automation system would integrate with its warehouse operations—both in the present and over a 5–10-year horizon. This modeling provided a strong foundation for building a business case that resonated with both leadership and staff. As Andrew Mackintosh, General Manager of Supply Chain at Douglas, explains, "The system is highly modular and easy to expand. If we need additional capacity, we can extend the grid without expanding the physical space. This approach has addressed our operational challenges and positioned us for future growth."

The deployment of intuitive and user-friendly technology also contributed to addressing labor shortages and improving workplace safety. The system reduced the need for manual lifting and eliminated worker interaction with machinery operating at height, as robots now handle these tasks. Employees now work at ergonomic stations focused on picking and packing, enhancing their roles rather than replacing them. Notably, the project did not lead to a reduction in permanent staff. Instead, workers were able to shift toward more engaging and higher-value activities. As one project leader noted, "Automation is a progressive step—it improves working conditions and enables employees to grow in their roles."

Aligned with its commitment to health and well-being, Douglas Pharmaceuticals views improved logistics and working conditions as extensions of its mission to "improve lives." Andrew adds, "This solution supports that mission by creating better working conditions for our people and helping us deliver products to customers more efficiently and accurately, ensuring critical health needs are met."

Meeting strict regulatory requirements, such as expiration dates and quality controls, is critical in the pharmaceutical sector. Automation helps ensure that products are delivered on time and in optimal condition. In addition to serving its own customers, Douglas also provides fulfillment services to third parties. The new system enhances flexibility and reliability for these clients by improving accuracy and traceability. Picking errors are automatically flagged, and products are handled less frequently, preserving the integrity of the packaging from warehouse to delivery.

Durability and energy efficiency were key considerations in the project. The bins and robots used in the system are designed for longevity, with some units still operational after nearly two decades in other installations. The energy footprint is minimal—10 robots consume less electricity than a household vacuum cleaner, thanks to smart charging strategies. This energy efficiency meant no upgrades were needed to the facility's existing power infrastructure, contributing to the project's overall cost-effectiveness.

The transition to automation was supported by strong internal communication and proactive change management. Douglas engaged its workforce early in the process to ensure alignment and a smooth implementation. As Andrew notes, "A significant amount of effort went into validating our assumptions through data and simulation. Each challenge we encountered became a learning opportunity. This was a collaborative effort involving the entire project team. As a company, we've always been forward-looking, and this project is a reflection of that mindset."

Automation systems such as the one implemented at Douglas are increasingly seen as a strategic tool across industries. Their scalability, efficiency, and relatively low implementation barriers make them well-suited for businesses aiming to optimize logistics while planning for long-term growth.

The automated storage and retrieval system from AutoStore has been tested and refined over nearly two decades, with ongoing enhancements contributing to greater efficiency and value in intralogistics. In recent years, several innovations have emerged to further streamline warehouse operations and support scalable automation strategies. Following are some notable advancements shaping the future of warehouse systems:

- **Starter Grid Solution:** A compact, modular, plug-and-play configuration designed for businesses at the beginning of their automation journey. This solution typically includes 1 6-meter frame, 4 robots, 4,000 bins, and 2 conveyor ports—delivering up to 99% picking accuracy and built-in scalability to accommodate future growth.
- **Intuitive Picking Assistant (IPA):** An operator guidance system that uses visual projection to support tasks such as picking and storage. The tool enhances accuracy, improves ergonomics, and can increase process speed by up to 50%, helping reduce human error while improving workflow.
- **Sensor Cleaning Station:** An automated cleaning mechanism that maintains optimal sensor performance on warehouse robots, minimizing downtime and preserving operational continuity over time.
- **No-Touch Pick and Pack Workstations:** A space-efficient setup that enables robots to pick items directly into two different carton sizes, seal them, and prepare them for shipment—reducing the need for manual intervention and accelerating fulfillment.
- **Automated Bin Induction:** A patented process designed to streamline and safeguard the initial induction of storage bins, making this stage of the operation faster and safer.
- **Warehouse Execution Platform (e.g., FulFillX):** A system that integrates with existing warehouse management software to optimize picking, packing, putaway, and inventory tracking. It provides real-time operational visibility through a single interface, accelerating system ramp-up and enhancing decision-making.

These advancements in automation—whether through Automated Storage and Retrieval Systems (ASRS), Autonomous Mobile Robots (AMRs), or other technologies—are becoming essential tools in modern warehouse and supply chain operations. They offer the flexibility, scalability, and resilience needed to address growing logistical challenges, labor constraints, and customer expectations.

Andrew Mackintosh of Douglas Pharmaceuticals advises, "Don't just solve today's problems—look ahead to future opportunities and consider where automation can take you."

As the logistics industry continues to evolve, ongoing innovation will play a critical role in shaping its future. The next wave of automation is expected to unlock even greater efficiencies, adaptability, and growth opportunities across a wide range of industries.

10. Summary

Warehouse automation systems are more widely implemented nowadays to improve the putaway, storage, and retrieval operations of pallets, picking of cartons, and product pieces. The automation would increase the storage density, resulting in much less warehouse space used. Reduced warehouse space gives significant savings to justify the business case. The improved capabilities, through automated operations, enable warehouse operators to be more productive, and companies can justify the improved salaries for warehouse staff. The automation can also enhance the orders' throughput to support the company's growing business, sometimes without the need to move the warehouse or increase the warehouse operations area.

The warehouse automation enhances the warehouse operations into Industry 4.0 execution, that is, digitalization of the material flow and workflows within the warehouse. The digitalization would lead to future intelligent warehouses which would use a higher level of automation, e.g. mobile arm robots, drones, and AI-enabled warehouse operations. The First step to achieving this vision is to "start small and think big" by forming a multi-skills team to study the automation project together. The next step is to get education and training on warehouse automation, and the team would need to review current operations and how to improve with warehouse automation. There are several small automation systems that can be easily deployed in the current warehouse environment, e.g. AutoStore, AMRs and AGVs, and sortation systems. The warehouse management and staff can get to learn to use these warehouse automation efficiently before moving on to more complex and high-tech automation.

Chapter 8

Cross-Docking and Lean Warehousing Practices: A Synergy For Efficiency

Learning Outcome

By the end of this topic, you should be able to do the following:

1. Analyze the role of efficient warehouse operations in enhancing customer satisfaction and maintaining competitive advantage.
2. Differentiate cross-docking from lean practices and explore their complementary role in optimizing warehouse processes.
3. Develop actionable strategies for integrating lean practices in warehouse operations using the DMAIC (Define, Measure, Analyze, Improve, Control).

1. Introduction

In today's competitive business environment, efficient and responsive warehouse operations are critical for achieving customer satisfaction and maintaining a competitive edge (Stock and Lambert, 2001). Warehouses have evolved from simple storage facilities to dynamic hubs, directly impacting a company's ability to deliver products quickly and cost-effectively. Rising customer expectations for faster delivery times and lower prices necessitate a continuous pursuit of improvement in ware-house efficiency (Chen *et al.*, 2013). This chapter examines two powerful tools for optimizing warehouse performance: cross-docking and lean warehousing practices.

We examine the theory behind each concept, analyze their synergistic relationship, and provide actionable strategies for implementation. This chapter also includes a practical case study to further illustrate the application of these principles in a real-world setting. By understanding and implementing these strategies, warehouse managers can transform their operations into well-oiled machines, ensuring a competitive edge and a superior customer experience.

2. Understanding Cross-Docking

Imagine a bustling airport terminal where passengers seamlessly transfer from one flight to another with minimal waiting. Cross-docking embodies a similar concept within logistics (Buijs *et al.*, 2016). It is a strategic approach that minimizes storage time within a warehouse by facilitating the direct transfer of goods from incoming deliveries to outbound shipments (Benrqya, 2019). Products are received, sorted, and allocated directly for onward journeys, bypassing extensive storage periods. This method streamlines the flow of materials, significantly reduces handling costs, and expedites order fulfillment, leading to several key benefits:

Reduced storage costs: By minimizing the time products spend in storage, cross-docking eliminates the need for vast warehouse space, leading to significant cost savings on rent, utilities, and maintenance (Ardakani and Fei, 2020).

Improved inventory management: With fewer products residing in the warehouse, inventory control becomes more manageable. This reduces the risk of obsolescence, overstocking, and stockouts, ultimately leading to a leaner and more efficient operation (Chen *et al.*, 2013).

Faster order fulfilment: Cross-docking eliminates the time-consuming process of putting goods away in storage and then picking them up later. This translates to faster order fulfillment times, enhancing customer satisfaction and potentially increasing order volume (Buijs *et al.*, 2016).

Enhanced supply chain agility: Cross-docking allows for a more responsive supply chain by enabling faster adaptation to fluctuations in demand. Products can be quickly directed to meet urgent customer needs without being bogged down in lengthy storage procedures (Benrqya, 2019).

Additional characteristics of cross-docking:
- **Minimized storage:** Products are typically held for a short duration, maximizing warehouse space utilization.
- **Efficient transfer:** Goods are sorted and allocated directly for outbound shipments, eliminating unnecessary put-away and picking activities.
- **Just-in-time (JIT) approach:** Cross-docking aligns with JIT principles, ensuring inventory arrives just before it is needed for outbound orders.

However, it is important to recognize that cross-docking is not a universally applicable solution. Here are some key considerations (Ladier and Alpan, 2016):

- **Demand predictability:** Successful cross-docking often relies on predictable demand patterns and coordinated communication between suppliers and customers. Unforeseen fluctuations in demand can disrupt the smooth flow of goods and create bottlenecks.
- **Supplier and customer coordination:** Effective communication and coordination with suppliers and customers are crucial for timely deliveries and order fulfillment. Disruptions in either area can negatively impact the efficiency of cross-docking operations.
- **Product suitability:** Not all products are suitable for cross-docking. Bulky or fragile items may require dedicated storage and handling, making them less suited to this approach.

3. Lean Warehousing Strategies

To implement lean strategies within your warehouse, the following steps can be taken:

- **Conduct a warehouse assessment:** Begin with a comprehensive assessment of your current operations. Analyze areas such as receiving, put-away, picking, packing, and shipping. Identify bottlenecks, inefficiencies, and non-value-added activities using tools, such as VSM (Bersamin *et al.*, 2015).

- **Design a streamlined layout:** Review your warehouse layout for optimal flow of goods. Minimize travel distances between receiving, storage, picking, and shipping areas. Create designated zones for picking frequently ordered items (Laosirihongthong *et al.*, 2018).
- **Standardize processes:** Develop SOPs for all warehouse activities to ensure consistency, reduce errors, and facilitate training (Prasetyawan *et al.*, 2020; Richards, 2017).
- **Implement kanban systems:** Use Kanban systems to manage inventory levels for frequently picked items. Replenishment signals can be visual cues, such as empty bins or designated reorder points (Buijs *et al.*, 2016; Cagliano *et al.*, 2023).
- **Invest in technology:** Consider implementing Warehouse Management Systems (WMSs) to automate tasks, improve inventory control, and optimize picking routes (Chen *et al.*, 2013).
- **Empower your workforce:** Train your employees on lean principles and involve them in continuous improvement initiatives. Encourage them to identify and suggest ways to eliminate waste and improve overall efficiency (Buijs *et al.*, 2016; Richards, 2017).

4. Integrating Cross-Docking with Lean Warehousing

Cross-docking and lean warehousing practices complement each other in enabling effective and efficient warehouse operations. These are detailed in the literature and summarized as follows (Chen *et al.*, 2013; Prasetyawan *et al.*, 2020; Tay and Aw, 2021):

- **Alignment with lean principles:** Lean principles such as minimizing storage time and optimizing product flow naturally align with the core concept of cross-docking.
- **Streamlined operations:** Implementing Kanban systems and optimizing picking routes further streamline the cross-docking process.
- **Efficient organization:** As emphasized by lean principles, a well-organized and standardized warehouse facilitates the efficient sorting and allocation of goods for outbound shipments in a cross-docking operation.

5. Case Study: Enhancing Warehouse Efficiency with Lean Six Sigma[1]

This section showcases a case study which revealed that space limitations, inadequate staging areas, and poor signage were significant contributors to inefficiency and safety concerns within the warehouse. The proposed solutions include implementing a 5S program, streamlined process flow, and clear signage for storage areas and standardizing inbound process steps, offering practical and actionable steps for Company G to enhance safety and efficiency within its warehouse.

- **Background of Case Study**

This case study addresses space-related challenges encountered by Company G, a third-party logistics provider, which have led to unsafe loading practices and improper item stacking within its warehouse. The primary aim is to enhance safety and efficiency by applying Lean Six Sigma methodologies and utilizing the Define-Measure-Analyze-Improve-Control (DMAIC) approach.

Company G specializes in warehouse operations and manages the client's warehouse, facilitating the movement of various items, including catering supplies (plates, cups, cutlery, service carts, etc.), materials used in airline seat assembly, documents, and food items, such as alcohol. The comprehensive management of the flow of goods within the warehouse encompasses both inbound and outbound processes, as depicted in Figure 1.

Due to the COVID-19 pandemic, Company G experienced a decrease in incoming shipments, resulting in surplus warehouse capacity. However,

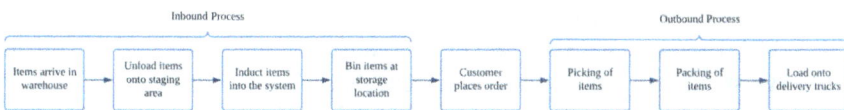

Figure 1. High-level process flow of the warehouse (adapted from Company G, 2023).

---

[1] This case study is partially based on a student's applied research project work on Company G. The author would like to acknowledge Francis Woon Li Yan for contributing to parts of the case study content.

as flight volumes regained momentum, there was a corresponding surge in incoming shipments, and the slow outflow of items led to space constraints within the warehouse. Consequently, cluttered items became a safety hazard, with insufficient space for safe loading, placement haphazardly, and improper stacking of items, compromising safety (Min, 2006).

The increased shipment volumes and the slow outflow of items have led to insufficient warehouse space, resulting in cluttered items and several associated issues. These issues include (1) insufficient space for safe loading, (2) placement of items in any available space, and (3) improper stacking of items (Staudt *et al.*, 2015).

Company G used Lean Six Sigma techniques to eliminate waste in the warehouse and provide solutions to improve efficiency and business performance. Company G eliminated waste in the warehouse and explored practical solutions to improve efficiency and overall business performance. The key tasks include understanding the current inbound process, identifying waste, streamlining the process flow, and proposing solutions based on Lean Six Sigma principles. The case study showcases how Lean Six Sigma is applied to address the challenges facing Company G through a structured approach. In particular, Company G carried out the following activities:

1. **Mapping and analyzing inbound logistics:**
   - to comprehensively map Company G's current inbound process flow, including receiving, inspection, and storage activities,
   - to identify and analyze the key performance indicators (KPIs) associated with the inbound process flow, such as receiving time, lead time, and accuracy.
2. **Applying lean six sigma to identify waste reduction opportunities and warehouse optimization:**
   - to leverage Lean Six Sigma methodologies to identify and eliminate waste within Company G's current inbound process flow,
   - to analyze the root causes of process inefficiencies and bottlenecks impacting inbound logistics,
   - to develop and evaluate potential improvement opportunities using Lean Six Sigma tools, such as Value Stream Mapping (VSM) and DMAIC (Define-Measure-Analyse-Improve-Control).
3. **Optimizing warehouse performance at company G:**
   - to identify and evaluate potential solutions for improving the overall efficiency and effectiveness of Company G's warehouse operations,

- to establish a comprehensive set of performance metrics to measure the success of implemented solutions, including order fulfillment speed, inventory accuracy, and warehouse space utilization,
- to assess the potential impact of identified solutions on Company G's overall supply chain performance and customer satisfaction.

The focus is on improving the inbound process of Company G's warehouse logistics chain to assist Company G in optimizing its warehouse operations by identifying and eliminating waste, improving efficiency, and proposing practical solutions aligned with the Lean Six Sigma methodology. This comprehensive approach is designed to enhance the company's overall business performance.

- **Lean six sigma for inbound logistics processes**

A comprehensive analysis of the current inbound process was conducted. The primary methodology adopted in this study for practical problem-solving and process improvement was DMAIC, also known as Define, Measure, Analyze, Improve, and Control, from the Lean Six Sigma methodology (Tay and Aw, 2021). The methodology included utilizing SIPOC diagrams, waste walks, time studies, and gathering feedback from staff to identify key issues and inefficiencies. In the logistics process, visual stream maps and fishbone diagrams were used to identify bottlenecks and root causes.

First, to understand the current inbound process at Company G, tools such as the SIPOC diagram and an As-Is VSM were used to look at the entire inbound process and capture the sequence of activities in the process. We conducted surveys, As-Is process maps, spaghetti diagrams, and time studies in the measure phase to identify waste and inefficiencies in Company G's inbound process, evaluate the performance of individual steps, and identify potential issues.

In the *analysis* phase, the collected data was analyzed, and root causes were validated using fishbone diagrams and time analyses. The "Improve" phase involved developing and testing solutions to address the identified issues, followed by implementing them in the control phase. The proposed solutions were evaluated for their effectiveness in improving the inbound process. Lastly, recommended solutions were proposed to support and enhance the current inbound process.

Overall, the DMAIC methodology and relevant Lean Six Sigma tools enabled the identification of inefficiencies in Company G's inbound process and the proposed improvement solutions.

In the ***measure*** phase of the DMAIC process, the primary data was obtained directly from Company G. The primary data collected consisted of warehouse observations to identify potential inefficiencies, as well as in-depth interviews with ground staff and managers. Throughout the study, a waste walk was conducted to identify any operational issues. All staff involved in the inbound process were interviewed, with questions ranging from "What are some of the issues you face at work?" to "What is hindering you from completing your work tasks?". These questions were asked to understand how people felt about the issues and outline the boundaries for the proposed solutions. It was essential that the ground staff feel comfortable with the solutions introduced. Secondary data from literature reviews was also considered due to the lack of information on using Lean Six Sigma to improve warehouses for 3PLs.

Quantitative data, specifically a time study, was also required for the time it takes for an item to be binned at its storage location from the time it is unloaded from the delivery trucks. The process was timed. Five samples were obtained and then averaged. Notably, Company G had not established any KPIs for the inbound process. The issues and challenges in the inbound processes were further complicated by a backlog of packages from previous years awaiting allocation to their storage locations.

In the analysis phase of the DMAIC process, the collected data was analyzed using the fishbone diagram and spaghetti diagram to understand the root causes of the issues. The data was merged with the As-Is VSM, and two kaizen burst areas were identified. Root cause analysis was utilized with a fishbone diagram to gain further insight into the issues.

In the ***define*** phase, the SIPOC diagram presented in Figure 2 was constructed to provide a high-level visualization of the inbound process in Company G while also understanding the inputs and outputs required. Following the SIPOC diagram, an As-Is VSM was developed to document each step of the inbound process, including the minor details executed in those steps.

Once the inbound process was clarified, the ***measurement*** phase started. As seen in Figure 3, an As-Is process map was also constructed.

Company G's existing inbound process starts with receiving goods from suppliers. Following the receipt, the goods are then unloaded. After that, the ground staff will conduct quality checks to ensure that items are correctly accounted for and in good condition upon arrival. If there are any discrepancies, Company G will inform its clients, while the remaining

Customer	Outputs	Process	Inputs	Suppliers
• Vendors • Suppliers	• Notification email	Goods arrive at Company B's warehouse	• Arrival of Goods	• Company B's Inbound Team
• Company B's Inbound Team	• MHE • Forklift	Goods are unloaded onto staging area	• Goods in Pallets	• Company B's Inbound Team
• Company B's Inbound Team	• Receipt from Vendor/Deliverymen	Goods are inspected and documents signed	• Shipment Documents • Receipt Received	• Company B's Inbound Team
• Company B's Inbound Team	• PO #, Part #, Batch #, Item description, Quantity	GR goods into SAP	• Update the quantity of goods in SAP	• Company B's Inbound Team
• Company B's Inbound Team	• GR-ed Item	Bin Goods	• Location	• Company B's Picking Team
• Company B's Inbound Team	• Location of Binned Item	Update Storage Location into PD Excel	• Storage location of goods are stored	• Company B's Picking Team

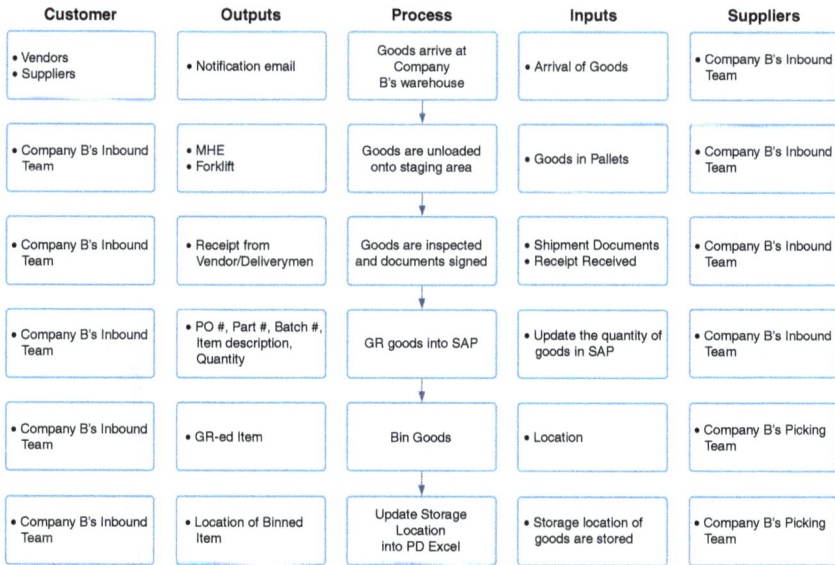

Figure 2.   SIPOC diagram of the inbound process.

goods are recorded in the system. Subsequently, the ground staff stores the items in their respective locations.

Waste walks were conducted throughout the study at Company G to better understand Company G's inbound process issues. Table 1 shows its consolidated analyses based on the Gemba walks and feedback from ground staff, summarizing the waste and improvement analysis conducted at the warehouse inbound logistics process. Table 1 combines the identified waste/issues within the inbound logistics process with the corresponding improvement strategies.

A time study analyzed the ground staff's movement. The time collected was integrated into the VSM, as seen in Figure 4. The As-Is VSM highlights each process's lead time (LT) and cycle time (CT). Lead time refers to the total time needed to complete a process, while cycle time denotes the time taken to execute a task. Kaizen bursts were placed at Process 3 and Process 5 due to the significant time disparity between the cycle time and lead time.

Based on the kaizen bursts identified in the As-Is VSM, fishbone diagrams were used to determine the root causes of Process 3 and Process 5.

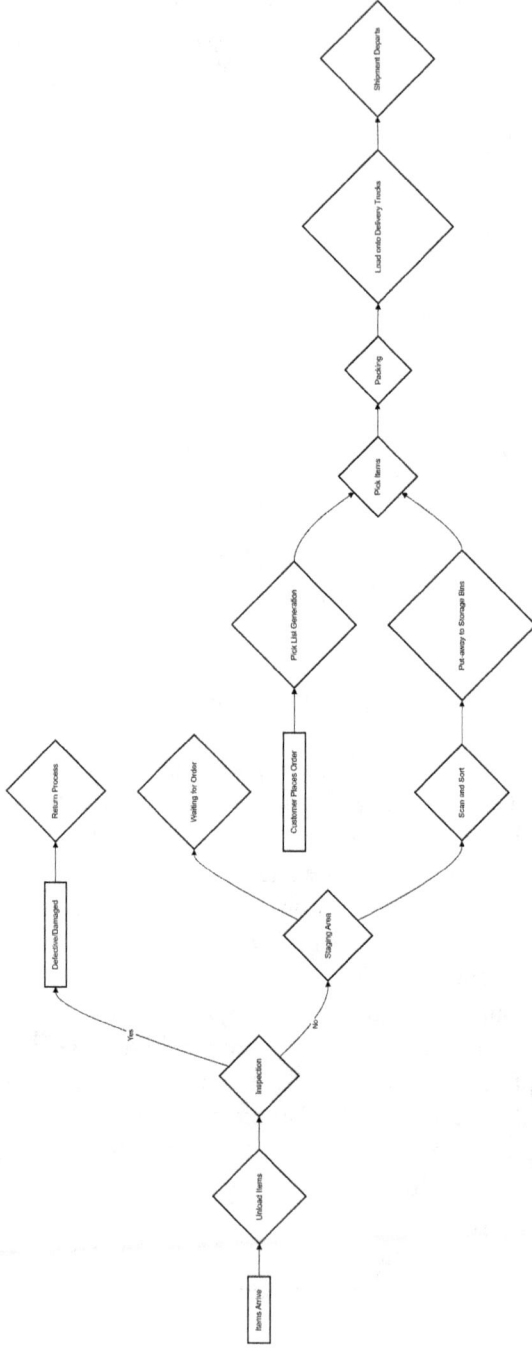

Figure 3. As-Is process map of the warehouse logistics process.

Table 1.   Waste identification and improvement strategy.

Waste/Issue	Description	Improvement Strategy
Transportation Waste	• Relocating goods due to insufficient storage. • Excessive movement searching for space.	• **Optimized storage:** Implement efficient storage strategies (designated zones and put-away optimization). • Utilize consolidation techniques for space maximization.
Inventory Waste	• Inefficient use of allocated storage space. • Waiting time due to unavailable inventory.	• **Inventory management:** Real-time inventory control systems. • Prioritize outbound fulfillment for efficient space utilization.
Movement Waste	• Time wasted walking to find storage space.	• **Flow optimization:** Strategic staging areas for efficient movement of goods. • Improved space allocation for easy access and reduced travel time.
Waiting Time Waste	• Delays caused by waiting for forklifts or space clearance.	• **Reduced wait time:** Scheduling forklift usage or alternative material handling solutions. • Streamlined processes to minimize waiting for space clearance.
Defect and Damage Waste	• Damaged goods due to improper storage location.	• **Quality control:** Designated inspection area for early detection of damaged goods. • Implement proper storage practices to prevent damage.

From Figures 5 and 6, the root causes of taking too much time were attributed to having to look for space in the warehouse and the lack of space to unload the goods. Furthermore, there were also staffing issues where they had to multi-task.

With the data obtained, issues discovered in the inbound process included a lack of space, a proper staging area, and no proper signage in the warehouse. Due to a lack of space, ground staff frequently had to maneuver goods in the warehouse. These items were either yet to be binned, picked, or placed there because there was no space in the warehouse or other levels. The spaghetti diagram further demonstrates repetitions in the individual process steps, another factor contributing to the inefficiency of the inbound process.

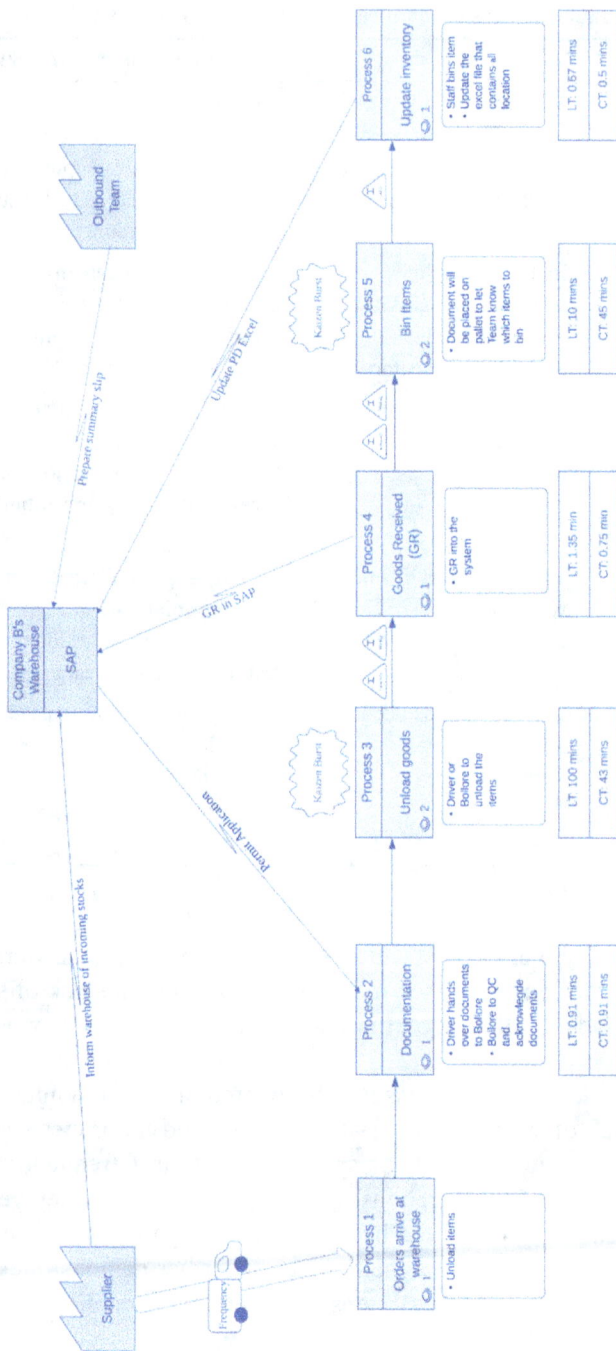

Figure 4.   As-Is VSM of inbound process.

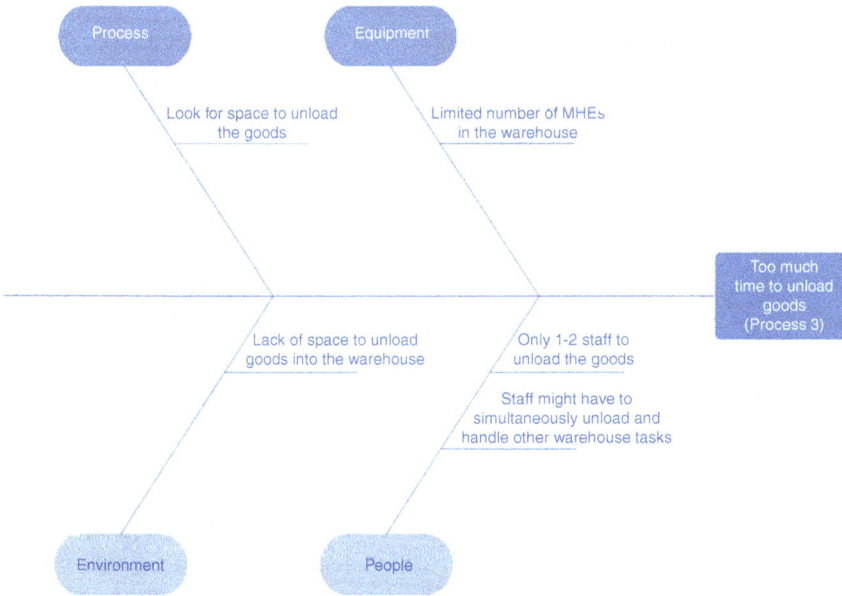

Figure 5. Fishbone diagram of process 3.

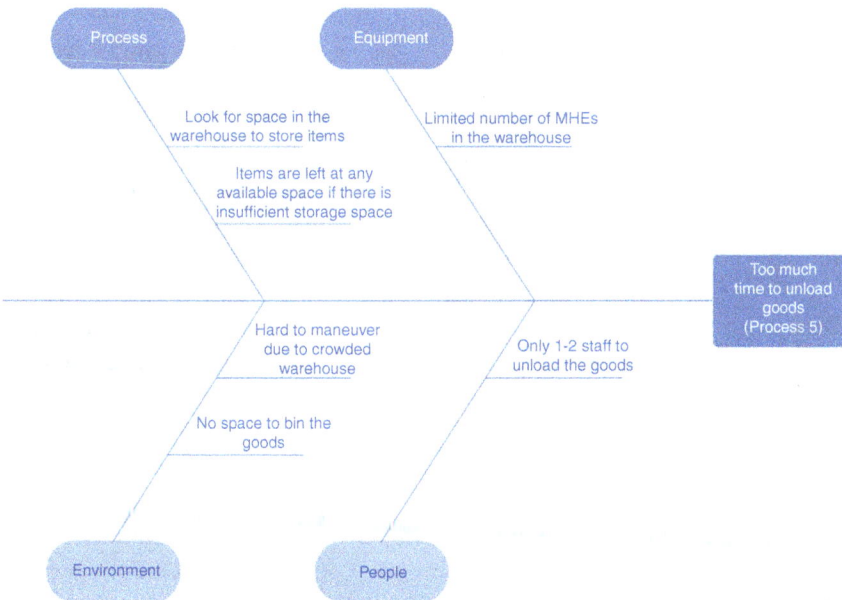

Figure 6. Fishbone diagram of process 5.

Proposed solutions:

Based on our understanding of the numerous issues and root causes identified in the warehouse, this section proposes solutions for Company G. The recommended course of action is implementing 5S: sort, straighten, shine, standardize, and sustain. The introduction of KPIs would also be helpful to Company G. Implementing these solutions would minimize inefficiencies and enhance its overall efficiency.

Sort: First, as many goods are lying around the warehouse, it is recommended to go through them and check whether they have expired or have discrepancies. These items should be relayed back to the client to see if they can be scrapped. It is encouraged that these items be placed together in another location to clear up some space in the warehouse, as this might take some time.

Straighten: To ensure that everything is organized, a lean physical flow is recommended. Additionally, a newly allocated staging area for inbound is located close to the loading dock. This ensures that items that have been unloaded can be easily brought into the warehouse while at the same time not blocking the pathway. This also makes space for trucks to fully unload the goods in the docking area.

Shine: In this step, signage and markings should be placed more legibly in the warehouse to ensure all goods are placed in designated places. For instance, the inbound and outbound staging areas should be demarcated so that staff will know what can be placed there and what cannot. The warehouse storage space can be clearly demarcated for the MHE's routes when moving around the warehouse. This ensures the safety of staff in the warehouse and guides the team when they move around to do their tasks.

Standardize: As not all inbound processes are standardized, the warehouse faced some issues. Only incoming overseas shipments are notified beforehand. Advanced email notifications should also be made for local and Malaysian shipments since there has already been a system in place for overseas shipments, whereby the client will email Company G with the required documents and a summary with the store consignment number, purchase order numbers, quantity, number of pallets, and dimensions of the pallets. Another proposed solution would be implementing dock appointment scheduling to support and better plan inbound logistical flow.

Table 2. Examples of KPIs that can be implemented.

Area	KPI (key performance indicators) — Time (hr) required for items to move to the next stage
Loading Bay to Inbound Staging Area	12 hrs
Staging Area to Binning	24 hrs

Drivers would have to schedule their arrival time with Company G. This will give the warehouse ample time to prepare for their arrival by creating space in the staging area. Staff can also schedule their tasks, reducing the risk of interruption and the need to attend to drivers and incoming goods while in the middle of their tasks. The required machinery can also be placed closer to the loading bay to prepare for the arrival of shipments.

Sustain: Finally, to sustain the 5S program, performance measures and reward systems can be implemented to measure the performance of each step of the inbound process. For example, KPIs can be implemented based on the time taken. Table 2 gives an example of KPIs that can be implemented.

Items in the loading bay should be brought into the staging area within 24 hours, while those in the staging area should be binned within 48 hours. This would prevent and reduce bottlenecks in the space and inbound processes, ensuring the constant movement of items.

Staff should also be rewarded for their hard work and ability to hit the proposed KPIs. This also ensures continued awareness of the program and encourages the team to do their best.

Case Study Questions

After studying cross-docking and lean warehousing practices and considering the real-world scenario presented in the case study, discuss how you would evaluate the potential implementation of these strategies in a warehouse operation.

1. Can you identify opportunities within the case study for incorporating cross-docking or lean practices?

2. Are there any challenges or limitations to consider when implementing these strategies in the specific context of the case study?
3. How might the successful application of cross-docking and lean practices impact the warehouse's efficiency, customer satisfaction, and overall business performance?

Acknowledgement

This case study is partially based on a student's applied research project work on Company G. The author would like to acknowledge Francis Woon Li Yan for contributing to parts of the case study content.

References

Ardakani, A. and Fei, J. (2020). A systematic literature review on uncertainties in cross-docking operations. *Modern Supply Chain Research and Applications*, 2(1), 2–22.

Benrqya, Y. (2019). Costs and benefits of using cross-docking in the retail supply chain: A case study of an FMCG company. *International Journal of Retail & Distribution Management*, 47(4), 412–432.

Bersamin, J., Drio, R., Lacibal, A.L., Manalastas, C., Mendoza, S., Orallo, G.M.D., and Tan, C.T. (2015). Design process using lean six sigma to reduce the receiving discrepancy report of ACE logistics. *Industrial Engineering, Management Science and Applications 2015*.

Buijs, P., Danhof, H.W., and Wortmann, J.C. (2016). Just-in-time retail distribution: A systems perspective on cross-docking. *Journal of Business Logistics*, 37(3), 213–230.

Cagliano, A.C., Zenezini, G., Rafele, C., Grimaldi, S., and Mangano, G. (2023). A design science–informed process for lean warehousing implementation. *IFIP International Conference on Advances in Production Management Systems*.

Chen, J.C., Cheng, C.-H., Huang, P.B., Wang, K.-J., Huang, C.-J., and Ting, T.-C. (2013). Warehouse management with lean and RFID application: A case study. *The International Journal of Advanced Manufacturing Technology*, 69, 531–542.

Ladier, A.-L., and Alpan, G. (2016). Cross-docking operations: Current research versus industry practice. *Omega*, 62, 145–162.

Laosirihongthong, T., Adebanjo, D., Samaranayake, P., Subramanian, N., and Boon-itt, S. (2018). Prioritizing warehouse performance measures in contemporary supply chains. *International Journal of Productivity and Performance Management*, 67(9), 1703–1726.

Min, H. (2006). The applications of warehouse management systems: An exploratory study. *International Journal of Logistics: Research and Applications*, 9(2), 111–126.

Prasetyawan, Y., Simanjuntak, A.K., Rifqy, N., and Auliya, L. (2020). Implementation of lean warehousing to improve warehouse performance of plastic packaging company. *IOP Conference Series: Materials Science and Engineering*.

Richards, G. (2017). *Warehouse Management: A Complete Guide to Improving Efficiency and Minimizing Costs in the Modern Warehouse*. Kogan Page Publishers.

Staudt, F.H., Alpan, G., Di Mascolo, M., and Rodriguez, C.M.T. (2015). Warehouse performance measurement: A literature review. *International Journal of Production Research*, 53(18), 5524–5544.

Stock, J.R. and Lambert, D.M. (2001). *Strategic Logistics Management* (Vol. 4). McGraw-Hill/Irwin Boston, MA.

Tay, H.L. and Aw, H.S. (2021). Improving logistics supplier selection process using lean six sigma — an action research case study. *Journal of Global Operations and Strategic Sourcing*, 14(2), 336–359.

Chapter 9

Warehouse Safety & Security

Learning Outcome

By the end of this topic, you should be able to do the following:

1. Identify common warehouse hazards, conduct risk assessments, and implement strategies to mitigate safety risks, ensuring a safe working environment.
2. Demonstrate effective emergency response protocols for fire, medical, and security incidents, and create an emergency plan to manage unforeseen situations.
3. Apply warehouse security measures to protect assets and ensure compliance with safety regulations and industry best practices.

1. Introduction

Safety and security are foundational elements of warehouse operations, working together to create a safe and efficient work environment. The primary goal of safety is to ensure that employees return home unharmed every day — a principle often highlighted by leaders in the industry.

Warehouses inherently pose numerous hazards due to the nature of their operations, making occupational safety vital for several reasons:

1. **Employee well-being:** Protecting employees' health and safety ensures they can work without fear of harm. For instance, Amazon

equips its workers with personal protective equipment (PPE) and conducts daily safety briefings to foster a safety-first culture. Fatigue-monitoring systems are also utilized to prevent accidents related to exhaustion.

2. **Legal compliance:** Adhering to safety standards is a legal requirement. Non-compliance can result in severe penalties, lawsuits, or operational shutdowns. In 2017, a UK warehouse operator was fined £400,000 for safety violations, underscoring the importance of following Health and Safety Executive (HSE) guidelines.

3. **Operational efficiency:** A safe workplace minimizes disruptions caused by accidents. Toyota's accident-free incentive program reduced workplace injuries by 25%, boosting productivity and traffic flow.

4. **Cost reduction:** Preventing accidents lowers expenses related to medical claims, legal fees, and damaged goods. For example, DHL's safety training initiatives led to a 30% reduction in workers' compensation claims, saving the company millions.

5. **Employee morale and retention:** A strong safety culture demonstrates that a company values its workforce, improving job satisfaction and retention. At XPO Logistics, employee involvement in safety committees has resulted in higher morale and lower turnover rates.

6. **Reputation management:** A robust safety record enhances a company's reputation, attracting clients and partners. FedEx, for instance, has consistently upheld high safety standards, bolstering its image as a reliable logistics provider.

However, poorly designed operations and inadequate training can undermine safety efforts, exposing employees to unnecessary risks. Globally, leading nations such as Singapore, the Netherlands, and Germany enforce strict safety regulations in high-risk industries, including construction, manufacturing, and logistics. Singapore's Workplace Safety and Health (WSH) framework exemplifies a balanced approach, combining stringent penalties with supportive initiatives, such as grants and training programs. With a target of reducing workplace fatalities to below 1 per 100,000 workers by 2028, Singapore has already achieved a record low rate of 0.99 in 2023.

Technology is another key enabler in enhancing safety. For example:

- **Vehicular safety:** Systems such as advanced driver assistance and 360-degree cameras mitigate risks from reckless driving or blind spots.

- **Construction safety:** IoT sensors and drones help prevent accidents, such as falls or collisions.

To sustain a safe work environment, companies must emphasize safety training, well-designed facilities, and regular equipment maintenance. New employees should undergo a comprehensive safety orientation covering topics, such as PPE, risk assessment, and emergency response. For existing employees, annual refresher courses reinforce safety principles. Additionally, providing ergonomic material handling equipment, such as forklifts and conveyor systems, helps prevent injuries. Warehouses must also prioritize layout design, including segregated pathways and adequate lighting, to ensure safe operations.

Regular inspections and maintenance of equipment further mitigate risks. For instance, forklift operators should perform daily checks on brakes, hydraulic systems, and signaling lights before commencing operations. In Singapore, forklift operators must complete a certification course and undergo on-the-job training (OJT) to ensure competence (see Appendix). By embedding safety into every aspect of warehouse operations — from training to technology — organizations can create environments that prioritize employee well-being, legal compliance, and operational excellence.

2. Fire Safety

An effective emergency preparedness and response plan for warehouses must include robust fire safety measures. These measures include the following:

1. **Fire extinguishers:**
   - Strategically placed in key areas, including exits and high-risk zones.
   - Compliance with local fire safety standards, such as the SCDF Fire Code in Singapore, which mandates a maximum travel distance of 20 meters to the nearest extinguisher.
2. **Sprinkler systems:**
   - Install automatic sprinkler systems per local codes, such as NFPA 13 or SCDF regulations.
   - Ensure in-rack sprinklers cover higher storage levels adequately.

3. **Fire detection systems:**
   - Use smoke and heat detectors connected to a central alarm system.
   - Ensure linkage to emergency services, such as SCDF, for swift response.
4. **Fire lanes and access:**
   - Maintain clear fire lanes for unhindered access by emergency personnel.
5. **Emergency lighting:**
   - Install adequate lighting to illuminate evacuation routes during power outages.
6. **Evacuation plan:**
   - Develop and communicate a clear evacuation plan, considering the warehouse layout.
7. **Regular inspections and maintenance:**
   - Conduct routine checks to ensure extinguishers, sprinklers, and alarms are operational.
8. **Employee training:**
   - Provide regular fire safety training, including extinguisher use and evacuation drills.
9. **Fire safety signage:**
   - Place clear and visible signs for fire exits, extinguishers, and assembly points.
10. **Fire drills:**
    - Conduct semi-annual fire drills as mandated by SCDF for premises with 10 or more occupants.

Adherence to these measures enhances warehouse safety, protecting employees, assets, and operations. Additionally, companies should address employee well-being through **Health and Wellness Programs**. Initiatives could include regular breaks, educational sessions on nutrition, and activities such as yoga or meditation. Offering perks such as free nutritious snacks can further support physical and mental health.

3. Security Protocols for Inventory Protection

Warehouse security safeguards physical assets, operational efficiency, and employee safety. Security challenges can be **internal** or **external**, with hardware and procedural controls addressing each.

3.1 *Internal security controls*

1. **Access control:**
    - Use keycards, biometric scanners, or security codes to restrict access.
    - Employ biometric systems for added protection against misuse.
2. **Secure storage:**
    - Lock high-value items in secure cages or storage areas.
    - Utilize intelligent lockers with RFID tracking for asset management.
3. **Inventory management systems:**
    - Implement barcodes, RFID tags, and Warehouse Management Software for accurate tracking.
    - Conduct regular cycle counts and annual stock-takes to detect discrepancies.
4. **Surveillance systems:**
    - Install CCTVs at strategic points like entry/exits and high-value zones.
    - Retain footage securely for at least three months.
5. **Security personnel:**
    - Employ trained staff to monitor activities, conduct patrols, and assist visitors.
    - Use guide dogs for nighttime patrols where necessary.
6. **Process control:**
    - Develop a Security Policy covering personnel monitoring, goods movement, and incident reporting.
    - Conduct thorough employee background checks, leveraging initiatives, such as Singapore's Yellow Ribbon Program for reintegrating ex-offenders.

3.2 *External security controls*

1. **Perimeter surveillance:**
    - Install fencing and night-vision cameras to monitor for unauthorized activity.
    - Integrate sensors with command-center alerts for intrusions.
2. **Visitor management:**
    - Implement badge or biometric systems for identifying external visitors.

- Ensure suppliers adhere to robust procedures for goods pickup and delivery, including verifying documentation and maintaining CCTV records.
3. **Loading bay controls:**
   - Use gate pass systems for vehicles, verified by security personnel.
   - Require acknowledgment from warehouse staff before departure to ensure proper handover.

4. Risk Management Strategy

Risk management is essential in warehouse operations to address hazards, ensure compliance, and improve efficiency.

4.1 *WSH framework and bizSAFE program*

In Singapore, the **bizSAFE Framework** helps companies enhance safety culture through five levels:

1. **Level 1:** Management commitment through WSH workshops.
2. **Level 2:** Risk management champion training and plan development.
3. **Level 3:** Implementation and audit of risk management plans.
4. **Level 4:** Development of a Workplace Safety and Health Management System (WSHMS).
5. **STAR Level:** Achieve SS506 certification for comprehensive WSH practices.

4.2 *Risk assessment process*

1. **Hazard identification:** Analyze physical, chemical, biological, and mental health risks.
2. **Risk evaluation:** Assess likelihood and severity using a 5×5 matrix.
3. **Control measures:** Prioritize elimination, engineering controls, and PPE.
4. **Documentation:** Maintain a risk register for tracking hazards and measures.
5. **Review and communication:** Regularly update and share findings with employees.

By involving key stakeholders, management, employees, auditors, and consultants, companies can meet safety standards, reduce incidents, and foster a safe working environment. In summary, safety and security are essential components of effective warehouse management. These two systems complement each other to minimize inventory loss and protect personnel. For instance, implementing robust access control measures, such as biometric scanners or key cards, ensures that only authorized personnel can enter the facility. This not only bolsters security by preventing unauthorized access but also safeguards employees from potential threats posed by intruders. Additionally, utilizing inventory management systems such as Warehouse Management Systems (WMS) that incorporate RFID or barcode technology significantly enhances operational efficiency while reinforcing security. By enabling real-time tracking of inventory movements, these systems help prevent theft and misplacement of goods, ultimately protecting both valuable assets and the employees responsible for handling them.

5. Case Study: Relocation of a Warehouse for a Cosmetic Company: A Case Study

Warehouses are pivotal components within a company's supply chain, handling crucial functions, such as receiving, storing, and delivering goods to end customers. While extensive research exists on the strategic location of warehouses in network optimization, there is a notable dearth of research on how to relocate a warehouse from one site to another, particularly while minimizing associated risks. This study sheds light on a low-risk approach undertaken by a multi-level marketing (MLM) company in Malaysia during the relocation of their warehouse. The study primarily addresses several key constraints:

(i) completion of the warehouse relocation exercise in 4.5 months,
(ii) termination of their warehouse staff services during the relocation,
(iii) elimination of disruption in the order fulfillment service level,
(iv) management of temperature-sensitive goods during the transfer and storage,
(v) seamless linkage between the third-party logistics (3PL) warehouse management system (WMS) and the company's systems.

The primary objective of this study is to propose a resilient strategy for the relocation of their warehouse, taking into account these key constraints faced by the company.

Relocating a warehouse is complex and time-consuming. It involves detailed planning, cost assessment, and risk management (Nagahan and Akın, 2018). Thus, effective project management is crucial to minimize stress, costs, and time overruns. Companies need competence in project management to improve supply chain processes. Different approaches, such as the Project Management Institute (PMI) approach or a more intuitive approach, can be used to enhance project management capabilities (Kuster *et al.*, 2023). Likewise, planning is vital in the warehouse relocation process. Factors such as trucking, labor, and equipment costs need to be budgeted carefully (Zapata *et al.*, 2020). Understanding the impact on a company's culture and stakeholder expectations is essential in managing risks (Singh, 2022). On the other hand, resilience is a dynamic process involving disturbance, surprise, change, and adaptation. Managing risk through a resilience perspective requires reducing, transferring, and preparing for impact, responding efficiently, and being prepared for unexpected events. Supply chain risk management is context-specific, focusing on critical and foreseeable risks (Gurtu and Johny, 2021). Hence, effective project management is essential, and a focus on resilience can enhance risk management strategies in supply chain processes (Ageron *et al.*, 2020).

This study, therefore, focuses on summarizing the development of a resilient approach for Company X to successfully relocate their warehouse. The primary reason for the study of a resilient warehouse relocation approach is that the consequence of a failure of warehouse relocation is very costly (Nagahan and Akın, 2018). Company X had two previous warehouse relocation projects that failed and suffered serious financial losses due to interrupted services. This is driven by the fact that Company X has not relocated its warehouse for more than 20 years and none of its existing supply chain staff has any warehouse relocation experience. Relocating a warehouse will always bring changes to existing proven systems and processes and hence introduce risks (Kotonen, 2017). Therefore, this study intends to propose a new structured risk identification framework and action plans to mitigate all the identified risks (González-Hernández, 2019).

This chapter is divided into six sections. It starts by briefly outlining the present Company X scenario and the major issues that arose during the relocation of the warehouse process. The current warehouse activities of

Company X are covered in the section. The risk assessment and mitigation approach, results, and KPIs are presented in the following section. The conclusion will then be presented in the final section. The results of this study will offer academic scholars and business practitioner's deeper understanding of warehouse relocation strategies. Additionally, the suggested strategies will improve supply chain risk management as a whole.

5.1 *Company X overview*

Company X is one of the pioneer companies in multi-level marketing (MLM) in Malaysia with over 40 years of history. It is one of the top MLM companies in beauty, household, and personal care categories with a global annual turnover amounting to billions of US dollars and millions of sales representatives selling their products to end consumers. It operates the same business model as its parent company in the US but with a smaller range of portfolio of products which includes skincare, color cosmetics, fragrance, and personal care products to suit local demands. As a direct sales company, Company X does not have any sales staff but instead recruits thousands of agents or distributors to sell their products directly to end consumers. To support these distributors, Company X operates more than a hundred retail shops across the nation to facilitate the replenishment needs of their distributors. In addition, it also offers online stores for distributors to place their orders even after office hours.

Company X was at its peak in 2004 with its share price trading at an all-time high of USD138.80 per share. However, Company X's performance deteriorated to the point where its shares were traded at only USD 4.08 per share in late 2018 or a negative 22.26% compound annual growth rate (CAGR) over these 14 years. The fall was due to mismanagement by the previous CEO, resulting in a decline of sales, profitability, and number of participating distributors. Consequently, a new CEO was hired in 2018 to turn around the company. The new CEO introduced a multi-pronged strategy that involved all levels of the company intending to improve its performance. This strategy included (i) sharpening the product portfolio, (ii) selling unused assets, (iii) shedding global workforce, (iv) hiring new global executives, and (v) renewing the company's focus on its biggest and best-selling brand. These strategies had significant implications on the supply chain, particularly items (ii) and (iii). Given these global directives from top management to operate with more agility and efficiency, Company X in Malaysia responded by selling off its HQ building (including the warehouse) in Petaling Jaya to a new buyer and outsourced its warehouse operations to a newly appointed third-party logistics (3PL) in a new location.

Company X sought out and appointed a 3PL to manage their warehouse operations from a different site. A pressing immediate concern related to staff, in particular, is making redundant the warehouse staff who average 12 years of experience. Compliance with local labor law required Company X to serve a 2-month notice or pay *in lieu* when dismissing any staff. Given their tight financial situation, Company X decided to serve their warehouse staff redundancy notice prior to the warehouse relocation exercise to reduce expenses. This however created huge operational risks such as sabotage by staff, in particular, absenteeism, intentional low productivity, delay of transfer plan, intentional errors in work such as picking the wrong item or quantity, damaging the goods, pilferages, etc. Additionally, local management has also set the requirement of no-service disruption to its sales agents from Monday to Friday during the move. This is to avoid a repetition of Company X's experience of two failed warehouse relocation projects with a loss of sales that ran into millions. Company X was 4.5 months into the completion of the relocation exercise when this study commenced. Thus, this study's objective is to propose a strategy for the timely and successful completion of the relocation exercise and ultimately prevent any delay in the handover of property to the new buyer the failure of which would expose Company X to legal penalties.

5.2 *Issues encountered*

In line with Company X's latest corporate strategies, the company decided to outsource its warehouse operations, sell off its warehouse building, and retrench its warehouse operators. However, moving a warehouse is not just about transferring stocks to another warehouse. It also involves the seamless relocation and continuity of the warehouse management system (WMS) and data, which will now be operated by a new team of warehouse operators from the 3PL. Furthermore, the company has to address the additional requirements which include a completion of the warehouse relocation exercise within 4.5 months. Company X sold off its existing warehouse and, as a result, needs to vacate and hand over the property in 4.5 months, failing which the company will incur costly penalties for non-compliance with the sales agreement. This means that all inventories, equipment, as well and racking facilities must be cleared from the current warehouse before the cleaning up and subsequent handover of the building.

Moreover, Company X timed the last working day of the warehouse operators to coincide with its planned warehouse handover date to maximize the savings on salary expenses. Thus, the notification of redundancy of warehouse operators before the commencement of the relocation exercise should be in place. According to Malaysian Labor Law, employers must serve a notice of termination to their employees in advance or must pay *in lieu* of notice for the same notice period. Company X chose to serve their warehouse operators 2 months' advance notice of the redundancy following labor law. Unfortunately, this means that the staff will be notified before any stocks are transferred out to the new warehouse. There would be pressure on the company to achieve a flawless relocation exercise. The greatest threat to a flawless relocation exercise in this project is the risk for the warehouse operators who are made redundant to either sabotage operations or lower their productivity.

Additionally, Company X also expects minimal or no disruption to its order fulfillment during the transition period as Company X is in an MLM industry and high service levels to customers offer them a competitive edge. Hence, product availability is critical to the business. Since some cosmetic items of Company X are temperature sensitive and require storage in a temperature-controlled environment, these items can only be transferred when the temperature control facilities are up and ready in the new warehouse. This adds to timing complexity in terms of planning and the transfer process. Moreover, Company X need not maintain its own WMS. The 3PL has the cost advantage of amortizing the WMS maintenance costs over a larger pool of users. However, the drawback here is the one-time integration cost between the two systems and the risks involved during the relocation. Company X had already appointed a new 3PL by the time the authors commenced research and proposal development on this project. The 3PL uses a different WMS from Company X which arose the need to develop, test, rework, and retest the system integration to ensure a smooth roll-out.

Managing any one of the above constraints on its own is not difficult. However, when taken together, they make the warehouse relocation exercise highly risky and prone to failure. This was the initial assessment of Company X's previous failed warehouse relocation project gleaned from available information presented about the company's action plan for those two previous exercises. It was clear that this project called for a low-risk approach to mitigate all possible risks that could arise from these five constraints on moving stocks, systems, and data from one warehouse to another.

For the record, Company X had recently performed two similar warehouse relocation exercises in markets outside of Malaysia. Unfortunately, both exercises ended up in failure and Company X suffered huge financial losses. The past projects used the Big Bang approach to transfer everything over one weekend and commence work on the new warehouse (system, operations, and products) on the following Monday. The authors were told errors happened since the go-live Monday and accumulated to such severity that they had to halt the new warehouse operations completely. The scenarios tested and passed during the user acceptance test with the EDI connection could not stand the test of the volume and frequency of actual sales orders. Staff was suspected of not supporting the warehouse relocation project wholeheartedly as it affected their livelihood. Unfortunately, the damage had already been done and they suffered lost sales amounting to millions of dollars and thousands of dissatisfied customers due to the unfulfilled sales orders. The problematic new warehouse stored 100% of its inventory and therefore no other warehouse could come to the rescue and share the burden of the unfulfilled sales orders. Eventually, Company X had to roll back the warehouse transfer and operate using their initial process and system.

5.3 *Methodology*

To develop a seamless and efficient warehouse relocation plan, this study first examines Company X's end-to-end supply chain in Malaysia. Geographically, Malaysia is divided into two regions by the South China Sea: West Malaysia and East Malaysia (Figure 1). Company X has its sole warehouse in Petaling Jaya which is the main warehouse that receives all incoming supplies and delivers the orders to all of their 100+ retail shops in West Malaysia and its two regional warehouses in Sabah and Sarawak. The regional warehouses in East Malaysia subsequently deliver the orders to the East Malaysian shops. The warehouse also fulfills all Internet orders nationwide and couriers the goods to the end customers (Figure 2).

Therefore, this study intends to propose a structured risk identification framework, as shown in Table 1, to capture all the key changes taking place in terms of area (i.e. physical, process, system, people, and vendor)

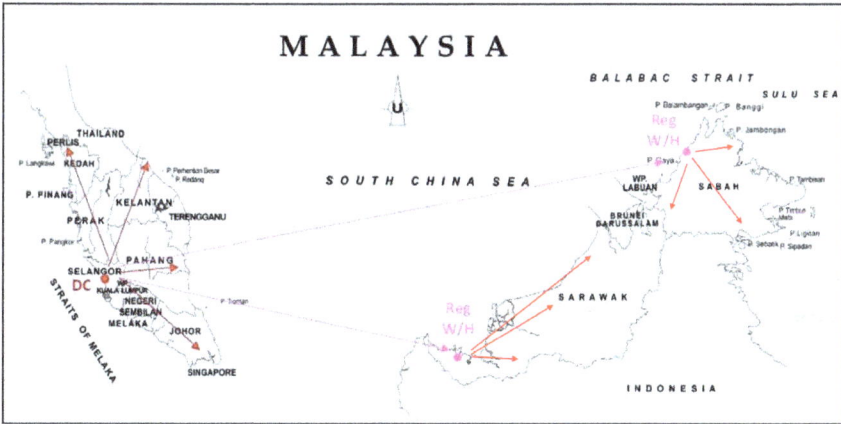

Figure 1. Company X regional warehouses.

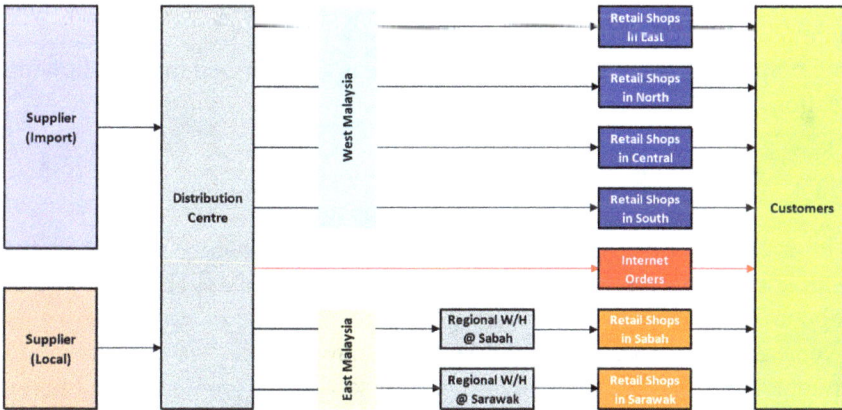

Figure 2. Company X's supply chain.

and stages of the transfer (i.e. at origin warehouse, during the one-time transfer or at destination warehouse).

5.4 *Analysis of the existing warehouse operations*

In order to develop a suitable solution for Company X, the authors must first understand its existing operations. The current warehouse operations

Table 1.   Warehouse relocation risk identification and mitigation framework.

Change Factors	Risks Identified			Mitigation Plan
	Origin Warehouse	One-Time Transfer	Destination Warehouse	
Physical				
Process				
System				
People				
Vendor				

at Company X cover several important areas as described in the following section.

5.4.1 *Receiving*

The receiving function is the starting point of inventory control in the warehouse. There are two types of incoming shipments. The first type of inbound shipment is the new or replenishment stocks ordered from either overseas or local suppliers. These stocks are shipped via container (sea freight from overseas suppliers) or truck (land from local suppliers). Company X receives on average 12 shipments per week; each shipment consists of roughly between 200 and 300 SKUs. The second type of incoming shipment is returned goods from end customers via Company X's agents and/or their retail outlets. An agent or retail outlet may return inventory due to multiple reasons, such as damaged products or incorrect product delivered in incorrect quantity or size. After the returned goods were received, the warehouse performed a quality check on the items to determine whether the product was fit for resale, in particular, for incorrectly sized items or incorrect items. If the product is deemed unfit for sale, it is then sent for disposal. On average, 98% of the inbound ships are the first type of inbound shipment and only 2% are the second type of inbound shipment.

5.4.2 *Organizing and storing*

All new and replenishment shipments are received in either a full pallet or in a secondary pack, depending on the minimum order quantity of each SKU. These shipments are then checked for correct quantity and condition before being stored in an assigned location on a pallet rack. These inventories then await two types of orders: sales orders or picking station replenishment orders. If there is a full carton sales order, it will be picked directly from the pallet racking area. If there is a picking station replenishment order, it will be picked and sent to the picking stations. At the picking stations, these cartons will then be broken into loose pieces to be put into the picking bins which are set up along the picking stations. A conveyor system runs through these picking stations, sending empty carton boxes with pick lists for operators to pick the correct item and quantity to be placed into the empty carton boxes. The carton boxes are then sent to the sealing station where they are sealed, and shipping labels are attached to every carton box. The sealed boxes are sent to the dispatch area to be sorted and then placed at the staging area according to the shipping lanes. Company X's inventory is categorized under 6 major product categories. These 6 categories consist of 8,102 different SKUs which take up around 6,889 pallet spaces. All inventories are assigned a status in the warehouse and Table 2 represents the breakdown of inventory by quantity and cubic volume based on the inventory status.

Table 2. Inventory breakdown by status.

Inventory	Active (items meant for sale and may be processed to fulfill a sales order)	Inactive (sale items but no transaction for the past 6 months)	Discontinued (sale items but no longer available for sale)	Non-Saleable (items not meant for sale, such as point-of-sales material)	Phased Out (items meant to be disposed of)	TOTAL
Quantity (pcs)	67,057,817	615,320	2,583,580	7,230,833	265,906	77,753,456
	86%	1%	3%	9%	0%	100%
Volume (m3)	8219	63	373	649	128	9,432
	87%	1%	4%	7%	1%	100%

5.4.3 *Order picking and packing*

Given that Company X's average sales order is about 30 line items with an average of 10 pieces per line item, predominantly in loose pieces, its picking area has set up a pick line with 1,200 bins on flow racks and a conveyor system to increase picking productivity. All sales orders received before the cut-off time are consolidated and released as a batch to the warehouse for their subsequent replenishment of the bins according to the batch of orders received before the next picking begins. There are two cut-off times for a day: (i) 8:30 a.m. for the morning batch run and (ii) 12:30 p.m. for the afternoon batch run. When the warehouse receives the orders, it replenishes the pick line according to the orders they received for the batch. Insufficiency of pick line inventory triggers a full carton replenishment order to the warehouse storage section for the required SKUs in full cartons. The pick section then receives these full cartons and breaks and replenishes the bins accordingly. Any unused loose pieces will remain in the carton and be sent back to the storage area in the picking section. Whenever an item is no longer needed in the pick bins, it will also be returned to the storage area of the pick section.

Once all the ordered items have been picked and placed into the shipping cartons, they are randomly checked for accuracy. Company X takes about 5% of the shipment for random checks. Subsequently, all shipping cartons are sealed and affixed with a shipping label displaying vital shipping information, such as (i) return address, (ii) destination address, (iii) recipient name, (iv) contact number, (v) package weight, (vi) shipping lane, and (vii) electronic tracking number and barcode.

5.4.4 *Delivery*

Company X processes and fulfills two types of sales orders and multiple modes of deliveries, as shown in Table 3.

Company X has more than a hundred retail outlets in West Malaysia. These outlets are grouped into 30 delivery lanes given their proximity within the same lane but different traveling distances and times for different lanes. Each outlet is filled with land transportation trucks, departing from Company X's warehouse around 7 p.m. 3 times a week. Company X also has about 20 retail outlets in East Malaysia which get their replenishment from their 2 local regional warehouses in East Malaysia: 1 in Kuching and 1 in Kota Kinabalu. Company X's warehouse only replenishes East Malaysia regional warehouses weekly via sea freight and not

Table 3.   Order type and delivery mode.

Order Type	Location	Recipient	Transportation Mode
Orders to replenish retail outlets	West Malaysia	Retail Outlets	Land—Truck
	East Malaysia	Regional Warehouse	Sea—Container
Orders to individual customers	West Malaysia	End User	Land—Courier Service
	West Malaysia	Staff	Land—Courier Service
	East Malaysia	End User	Land—Courier Service

directly to the retail outlet. Company X also sends orders directly to its end users who place orders online. Their online orders are processed on the same day if received before the processing cut-off time or else processed the following day. To get the orders promptly to the end users, Company X uses the services of a courier company that collects the packages between 5:00 p.m. and 7:00 p.m. daily. Staff orders are also similarly shipped through a courier company following the same process as end users. Company X warehouse also includes online orders from Singapore for the reason of economies of scale. It is the same process as the Malaysian online orders, except Singapore orders have one additional process: custom clearance in Singapore. As such, these orders are handled by a different courier company possessing the expertise to clear Singapore customs.

5.4.5 *Delivery lead time*

Given the traveling distance, all the collected land orders are delivered the next working day, including orders to Singapore. However, sea order delivery to East Malaysia is based on the vessel schedule. A container is brought into the warehouse for stuffing and then sent out to the port a few days before the vessel's departure time.

5.4.6 *IT system*

Company X's IT setup is relatively simple: OMS to handle orders from multi-channels, ERP to process the incoming sales orders and handle other enterprise transactions (e.g. purchase orders), and WMS to control the operations within the warehouse. Figure 3 demonstrates the high-level illustration of Company X's current IT setup.

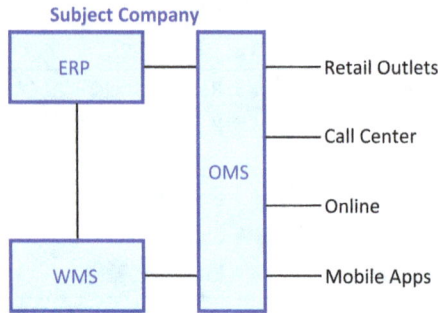

Figure 3.   Existing IT setup.

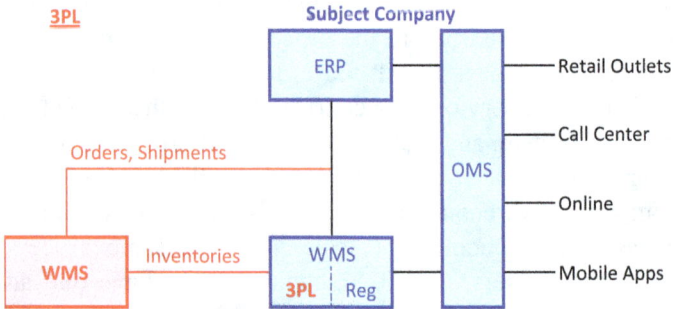

Figure 4.   New IT setup.

Moving forward, when a 3PL uses a different WMS from Company X, the IT architecture landscape needs to change. Whenever there is change, there is risk. The to-be IT architecture will have to link with the 3PL's WMS for the timely transmission of information. Further Company X still has its regional warehouses in East Malaysia to maintain. Thus, a new warehouse location "3PL mirror" has to be created in Company X's existing WMS, and Application Programming Interface (API) needs to be created to link 3PL's WMS back into Company X ERP and WMS, as shown in Figure 4.

5.5 *Findings and discussion*

5.5.1 *Risk assessment of relocation exercise*

To systematically identify all the risks in this study, the authors look for changes that will occur in all three stages of the transfer, as shown in

Figure 5: Stage 1 at Origin Warehouse (Company X's warehouse), Stage 2 during the one-time stock transfer process (from Company X to 3PL), and Stage 3 at Destination Warehouse (3PL's new warehouse).

Subsequently, at each stage, all the activities are scanned through for any departure or variance from current practices to identify what exactly will be changed and their potential risks. Then, for every change identified, the authors developed the potential risk associated as well as the probability of occurrence, as shown in Table 4.

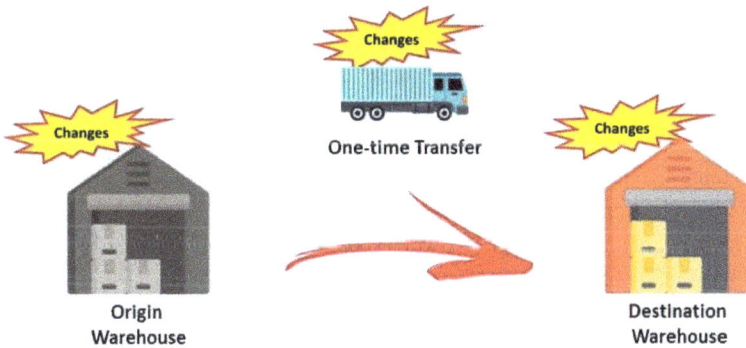

Figure 5. The 3 stages of transfer.

Table 4. Warehouse relocation changes and risk identification template.

| Change Factors | Transfer Stage | | | Risk | Profitability |
	Origin Warehouse	One-Time Transfer Process	Destination Warehouse		
Physical					
Process					
System					
People					
Vendor					

The authors would then record all the change factors identified and the risks involved, as summarized in Table 5.

Finally, the authors developed a mitigation plan for every change and risk identified from the exercise above. To ensure success, Plan B is developed on top of the initial mitigation plan so that the authors would know exactly what to do if the authors encounter failure on launch day, as depicted in Table 6.

Table 5.   Changes and risk identification.

Factors	Risk	Mitigation
Capacity	Is the storage capacity sufficient for the next 2 years of Company X's business?	New warehouse capacity is designed to meet in the new warehouse.
Storage Condition	Does the new warehouse have the appropriate temperature storage condition for Company X?	No new products with different temperature storing conditions. The new warehouse is designed to cater to existing temperature requirements.
Equipment	Does the new warehouse have the right type and sufficient Material Handling Equipment (MHE) fleet for Company X's operations?	The same storage and operating process is in the destination warehouse as the origin warehouse. Thus, the same MHE is adopted but the quantity is based on the next 2 years' demand.
Utilities	Is the electrical/water/broadband/ Wi-Fi supply in the new warehouse reliable?	The new warehouse already has a standby power generator built and a temporary water supply from a tanker has been arranged. Extra Wi-Fi dongles are bought and available should broadband fail one day.
Operation readiness	Are the new warehouse operation procedures suitable for Company X's requirements, in terms of volume, speed, and accuracy?	Jointly develop the new SOP together with Company X and 3PL so that Company X's insight can be transferred to 3PL.

Table 5. (*Continued*)

Factors	Risk	Mitigation
Transfer	Company X had never moved a warehouse before. Massive losses in inventories could occur during the transition if they are unaccounted for by both parties. The transfer workload is over and above the existing staff's workload. Not all transfer tasks could be completed in time.	Commence the transfer early, prepare to work overtime or during the weekend if the transfer is behind schedule, and increase resources to speed up the transfer.
IT System	System integration does not work properly when going live. System integration fails persistently after it goes live.	Set up a cross-function hypercare team to be on standby from go-live onwards to fix any occurring issues immediately. Implement Plan B. Roll the new operations back to the origin warehouse where the existing process and system are proven.
People	Company X's staff would be notified of the redundancy notice before the warehouse transfer and they may sabotage Company X's transfer operations by stealing the stocks or purposely mixing the wrong items in the shipping carton, working slowly, or having poor attendance.	Compensate the staff following local labor law for their years of service with the company. Maintain constant dialog with affected staff and arrange job opportunities for them in 3PL. Have another backup plan for labor resources from 3PL to come to Company X and operate the warehouse.

5.5.1.1 Relocation approach and assessment

To migrate the warehousing operations to the 3PL without any disruption to current operations and service levels, three relocation approaches were evaluated. Based on Company X's failure of two previous relocation projects causing extensive losses, a major process management change in the way the relocation exercise was implemented and executed was critical in

Table 6.   Risk mitigation action plan.

Change Factors		Transfer Stage					
		Origin Warehouse	One-Time Transfer Process	Destination Warehouse	Risk	Profitability	Mitigation Plan	Mitigation Plan Strategies
1. Physical	Warehouse Capacity			Different building	Wrong setup and insufficient capacity	Low	Design for the next 2 years demand	Source for overflow warehouse
	Storage Temperature			Different building	Wrong setup and insufficient capacity	Low	Design for next 2 years' demand and new product development	Source for overflow warehouse
	Equipment			Different setup	Wrong setup and insufficient capacity	Low	Follow the same type. Back up from 3PL other clients	Lease more
	Utilities			Different building	Failures	Low	Power generator. Chartered water tanker. Extra wi-fi dongles	Get third site on standby

Category	Subject	Source	Risk	Severity	Mitigation	Continuous improvement
2. Process	Warehouse Operations	New 3PL's processes	Missed functionalities and productivity	Mid	Develop 3PL's SOPs together for knowledge transfer	Continuous improvement
	On-Time Stock Transfer	Workload, speed, and accuracy	Missing, wrong, dispute, and delay on transfer	High	Joint count with 3PL at each level and start traceably	Work overtime and weekends
3. System	PL's WMS	Different WMS	Missed functionalities.	Mid	Rigorous tests: unit test, user acceptance test, endurance test, and stress test Hypercare team on standby	Go live early but on a small scale progressively
	Integration with 3PL	Different WMS	Failed integration	High		
4. People	Subject Company's Staff	Made redundant	Attendance, productivity, and sabotage	Mid	Lawful compensation, open dialog and exit assistance	3PL crew to take the origin warehouse operations
	PL's Staff	New crew	Mistakes and productivity	Mid	100% scan on pick	Start early to go up the learning curve

(Continued)

Table 6.   *(Continued)*

| Change Factors | | Transfer Stage | | | | | | |
| | | | One-Time | | | | | Mitigation |
		Origin Warehouse	Transfer Process	Destination Warehouse	Risk	Profitability	Mitigation Plan	Plan Strategies
5. Vendor	Suppliers			New location and processes	Mistake and delay	Low	Assistance at the new site for familiarization	
	Transporters			New location and processes	Mistake and delay	Low	Same transporters. Assistance at a new site for familiarization	3PL's transporters as standby

ensuring that history did not repeat itself. In order to do so, an approach based on a business reengineering process was used and three approaches were evaluated to determine the new approach to adopt. The three approaches were deliberated next.

5.5.1.2 Big Bang Approach: 100% inventories, all over a weekend

In this approach, 100% of the inventory in the existing warehouse is moved to the new 3PL warehouse over the weekend. The 3PL then receives these inventories into their own WMS and immediately begins operating 100% of the functionalities from Monday onwards. Company X's existing warehouse then ceases operations from the same Monday onwards as all stocks are already in 3PL's new warehouse. Any orders after that transfer cutoff deadline are fulfilled out of the new 3PL warehouse. Using this approach, only one warehouse is fully operational at any moment in time. Figure 6 illustrates the Big Bang approach.

The advantages of this approach are a cleaner relocation process with a clear cut-off timeline for the transfer. It is easier for staff, operators, and customers to comprehend and remember the day of the relocation to be

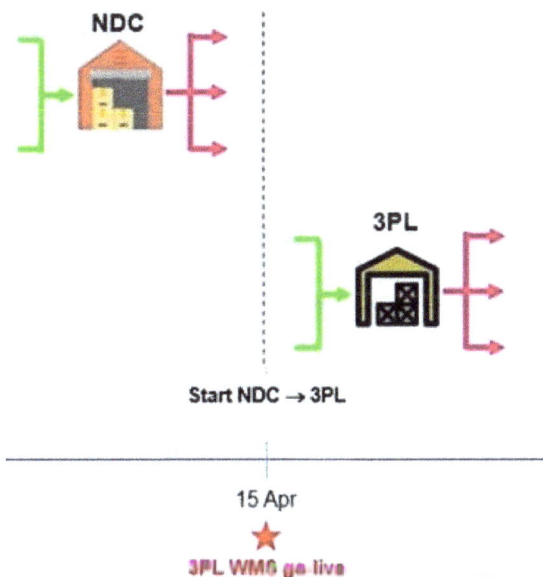

Figure 6.   Big bang approach.

known as "D-Day". Additionally, there is no need to maintain multiple systems, processes, or teams of people as opposed to the parallel run approaches and lower operating costs for the 3PL as 100% of their resources could be planned and commenced to work from that one day. Nevertheless, the disadvantages of this approach are considered high risk as the "placing all eggs in one basket" approach. If there are any unforeseen errors, then operations for 100% of the inventories are at risk. Additionally, this approach needs a huge fleet of trucks and manpower to load 6,889 pallets onto trucks, drive 6,889 pallets over to the 3PL warehouse (or 345 trips), unload 6,889 pallets, count and receive 6,889 pallets into 3PL WMS, put away 6,889 pallets into the correct storage location, fill up 1,200 picking bins with the correct items, etc., all over a short time span of a weekend.

In the Big Bang approach, only one key advantage was identified. It involved the reduction of the risk of an out-of-stock situation created when the inventories of all SKUs are required to be split into two different warehouses resulting in reduced stock coverage at each warehouse. By fulfilling orders from only one warehouse at a time, inventory can be pooled to increase the stock coverage of a certain SKU. The disadvantage of this is that the success of the warehouse relocation exercise is solely dependent on the 3PL service provider's speed and efficiency in transferring and receiving both the physical stock and data into their WMS system. Any delay results in orders not being fulfilled within the expected service level lead time which results in poor customer satisfaction and loss of sales. Further, this approach does not offer any risk mitigation alternatives to ensure that there is no supply disruption. Service levels are crucial for an MLM company in ensuring that the relationships between customers and distributors are not jeopardized. Thus, the authors do not recommend this approach due to its high-risk element as well as the high concentration of resources required over a weekend which is unfeasible for Company X.

5.6 *Gradual transfer approach: 100% of product category, category by category*

The second approach involves a gradual transfer of inventory, product category by product category from Company X warehouse to 3PL. For example, in Week 1, Company X would transfer all skincare products over to the 3PL new warehouse and subsequently transfer 100% of another product category in Week 2. Using this approach, 100% of stocks of the selected product categories are moved to the 3PL warehouse. The decision on which warehouse fulfills an order is based on where the inventory of SKUs in a certain product category is located. For example, skincare care orders are fulfilled from the 3PL warehouse in Week 1 while other product category orders are fulfilled from Company X's warehouse. Figure 7 illustrates the gradual transfer approach by product category. This means that when a sales order consisting of multiple product categories is received, some of the line items are fulfilled from the 3PL new warehouse while the rest are fulfilled from Company X's warehouse. Using this example, the customer service staff is then required to route the skincare order line items to 3PL while other order line items to the company. Company X's existing warehouses are based on where the inventory is located at that point in time. The same SKUs will not be located at both warehouses concurrently and only one warehouse is able to fulfill an SKU at any moment in time.

The advantages of this approach eased off the huge fleet and resources required for the Big Bang approach. The transfer can be stretched over

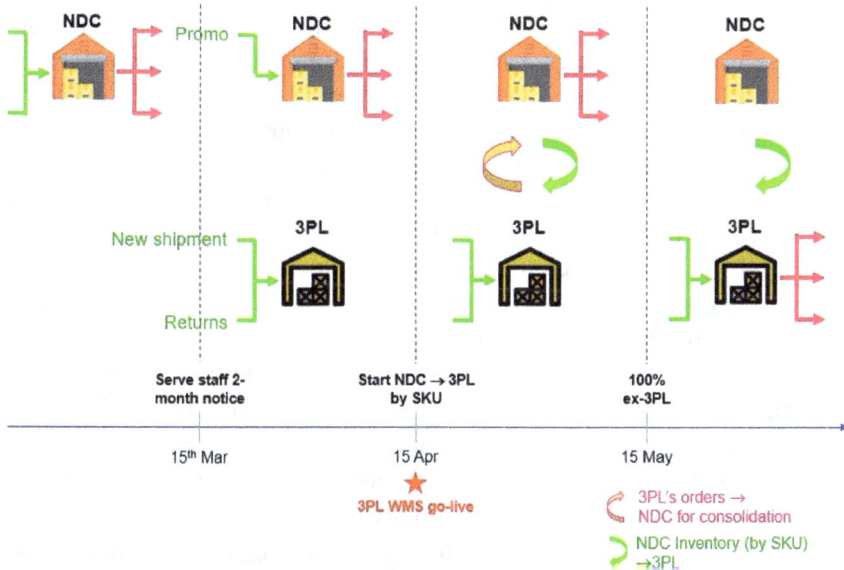

Figure 7. Gradual transfer approach by product category.

several weeks whereby the fleet size and resources would not result in a bottleneck. This approach also offers lower risk. At any one transfer, only one product category is involved, and therefore should any issue occur, only that category of products is at risk. However, the disadvantages of this approach are costlier where two operations teams and systems need to be up and running during the overlap period. This approach also creates additional complexity and attendant problems given that there are two fulfilment points for sales orders involving multiple product categories. This can be resolved by consolidating the products from two warehouses into one carton box before delivery to end customers. However, this causes additional processes and increased costs. Additionally, there are higher chances for delay and errors during the consolidation of products from two warehouses. Thus, do not consolidate the products but send two separate shipments from two different warehouses for one sales order. This creates confusion among the end users as they would be uncertain whether the first carton received is incomplete because of two fulfillment sources or a wrong pick by the warehouse.

Although the relocation risk has been reduced by this approach, the authors do not recommend this given its additional and higher complexity in accurately merging two separate orders with any system or tool support which is very risky. Furthermore, creating confusion among end users is also not an acceptable approach. However, relocation of order fulfillment and inventory by-products has several advantages. The first advantage allows for the early transfer of goods to the 3PL warehouse to move up the learning curve earlier. This allows the 3PL warehouse staff more time to familiarize themselves with the different products and new processes of receiving, staging, put-away, storage, issuance, picking, packaging, and shipment. Second, since order fulfillment for certain products is carried out by only one warehouse at a time, there is no need to split the inventory and hold it at two different locations. This results in a higher stock coverage for each product thus reducing the probability of the occurrence of a stock situation. This approach, however, creates additional complexity in the order fulfillment process as now one order could involve products that are located in two different warehouses (existing warehouse and 3PL warehouse). Additional work is also created as either the 3PL warehouse needs to pick and fulfill the order partially and then shunt it to the existing warehouse for further consolidation before shipping, or vice versa. Apart from the additional cost incurred from shunting, where the orders are not consolidated and shipped separately from different warehouses may cause an increase in transport costs which would offset any savings from the warehouse relocation exercise. Apart from introducing additional handling of goods, this approach is also expected to increase the order fulfillment lead time due to the additional steps in the fulfillment process that may result in damaged or misplaced goods.

5.7 *Gradual transfer approach: 100% of a region volume, region by region*

The third approach is to transfer 100% of a region's forecasted volume to a 3PL new warehouse, one region at a time. Instead of moving 100% of a product category (at a fraction of all SKUs), the authors transfer 100% of all SKUs at a fraction of the national volume (based on regional sales forecast volume) to the 3PL new warehouse. Using this approach, the authors attain the benefits of both worlds, minimization of risks arising from the gradual transfer (only a fraction of all inventories is at risk at any single moment in time), as well as the elimination of additional complexity from consolidating fulfillments from two different warehouses into one delivery. In this approach, a portion of the inventory of all active SKUs is transferred to the 3PL warehouse. This results in both the existing warehouse and the 3PL warehouse carrying the same type of SKUs at the same time. This will allow any orders to be fulfilled by either one of these warehouses. However, the decision on which orders will be fulfilled by which warehouse is made is based on the region the customer is located in. When customer service staff receives an order, they are to route it to the correct warehouse based on the ship-to address. Figure 8 illustrates the gradual transfer by the delivery region.

The advantages of this approach are minimal risk given it is a gradual transfer as well as avoiding the need to consolidate orders from two warehouses because all SKUs are available at both warehouses at any moment in time. There is also a partial inventory move of all active SKUs. Similar to the second approach, the third approach of fulfillment by region/customer offers the same advantage of speeding up the learning curve. Any risk associated with the possibility of sabotage from the existing warehouse staff who may be disgruntled after being served with redundancy notices will also be reduced as a turnkey backup plan is available should any issues arise. Additionally, for whatever reason should the existing warehouse suddenly experience a major disruption to its operations, the

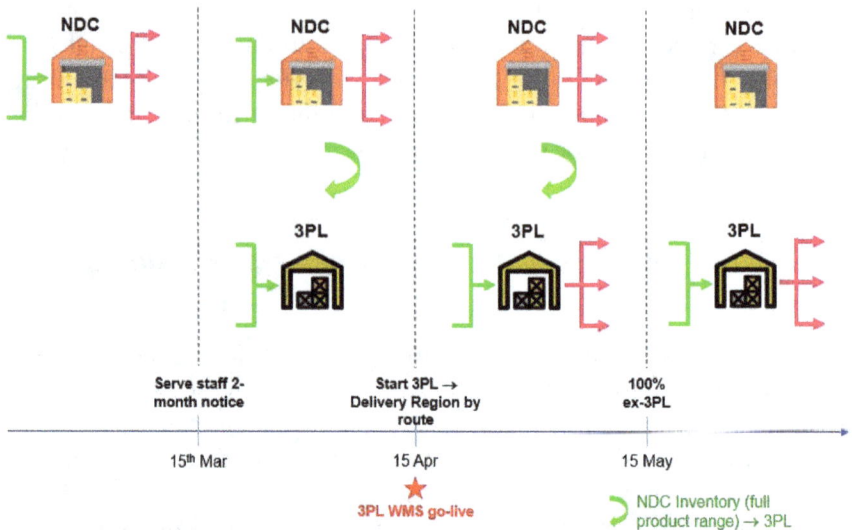

Figure 8. Gradual transfer approach by delivery region.

3PL warehouse is able to function to fulfill the orders. Lastly, there is no shunting of partial orders, and matching is required as only one order can be fulfilled by either warehouse. This will also eliminate the requirement for additional order fulfillment lead time and increase transportation costs. However, the risk of an out-of-stock brought on from reduced stock coverage is increased in this approach, especially for SKUs that are low in physical quantity as it becomes even harder to meet the demand. For example, the nationwide demand for SKU A is 10 pieces and the existing warehouse has 15 pieces on hand. If during the relocation process period only 5 pieces are transferred to the 3PL warehouse, but an order was received for 7 pieces, this would result in an out-of-stock which would negatively impact service levels to the customer.

5.8 *Approach to implement*

A decision on which approach to adopt during the implementation of the warehouse relocation exercise was based on considering a few factors including a thorough assessment of the advantages and disadvantages of each approach. Additionally, the risk appetite of Company X taking into consideration all the risks identified during possible risk mitigation strategies and the business requirements of Company X in terms of service levels and disruption to its existing operations. After careful consideration and deliberation with the management of Company X, it was decided that a "Gradual Transfer — 100% of a region volume, region by region" was to be adopted as it offered the lowest risk impact out of all the three proposed approaches. The transfer approach was then implemented, and the success of the transfer strategy was measured based on the results of a series of Key Performance Indicators (KPIs).

5.9 Results and KPIs

The purpose of a warehouse relocation project is to transfer the functionalities from the origin warehouse to the destination warehouse. Therefore, the yardstick on whether a warehouse relocation project is successful or otherwise depends on the success of the new operations in the destination warehouse. Since the key functionalities of a warehouse are goods receiving, storage, and delivering the goods to end customers, the authors set 1 system integration and 3 operation KPIs to ascertain the success of our warehouse relocation strategy which is deliberated in the following section.

5.10 System integration with the 3PL company: Order transmission

The creation of customer sales orders, both before and after the warehouse relocation sent to Company X, remains unaffected. However, changes do occur after the orders are received in Company X's order management system (OMS). In the new arrangement, the sales orders must now be sent to the 3PL warehouse, instead of the Company X warehouse. Given that the 3PL uses a different WMS, the authors used an API, a software intermediary that allows two applications to talk to one another. Sales orders are now being electronically transmitted from Company X's OMS to 3PL's WMS with immediate effect via this API. However, API is new to Company X and therefore it carries potential risks including (i) design flaws, wrong architecture (data points are connected wrongly), and incomplete architecture (some data points are not connected), (ii) speed (orders take too long to reach 3PL WMS), (iii) capacity (cannot handle large enough volume), and (iv) timeliness (the transmission interval is too long).

Although the authors had taken pains during the unit test, user acceptance test, and stress test to ensure the API was properly developed, nothing compares to actual live performance. The sales orders need to be approved by Company X first before they can be transmitted over to 3PL for processing. This is the same process as before and the manual order approval process is maintained. To assess the order transmission performance, the authors measured the time of the batch of orders placed in the morning and noon that reached 3PL WMS. Should there be any flaw in the API development, this would be captured and would appear negatively in this KPI. Before the relocation, Company X had two batches of orders released to the warehouse for processing every day: one in the morning and one around noon. Orders were being released in batches and not real time for higher productivity reasons, as the warehouse needed to replenish the pick lines according to the items ordered before the batch run could begin. This warehouse relocation has now added a new API process to link the Company X system to 3PL.

The authors also tracked the timeliness of the orders being received by 3PL at 8:00 a.m. and 12:00 noon from Day 1 of the go-live onwards. The authors observed a total of five delays over the one-month period— three of which were due to system bugs, while the other two were due to intentional human intervention. All three system failures occurred on the first day of a new phase, e.g. Go-live (East), the addition of North, and the addition of Central. This is because all changes present inherent risks. The progressive transfer strategy is to divide the risks into smaller manageable sizes over a longer period so that the team can divide and conquer one by one. Even if the issue occurs, the impact is only on a fraction of the total business volume that would be impacted, i.e. 20%, 45%, 70%, 99%, and 100% in this case. Figure 9 shows the timing of orders received by the 3PL WMS, while Figure 10 shows the impact of the timing of orders received on business volume. Next is to assess the success of the original warehouse operations that have been transferred to the destination warehouse. The authors set up three KPIs to measure its picking, packing, and delivering activities, respectively.

5.11 *Operations: Picking productivity*

Once the orders are successfully transmitted over to the 3PL warehouse, the operations team begins to print out the sales orders to be inserted the orders into the shipping cartons accordingly. Then, pickers pick the

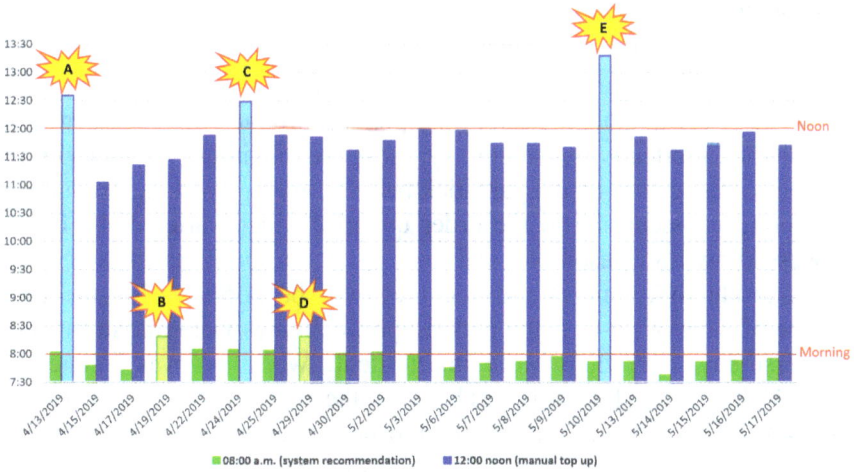

Figure 9. Timing of orders received by 3PL WMS.

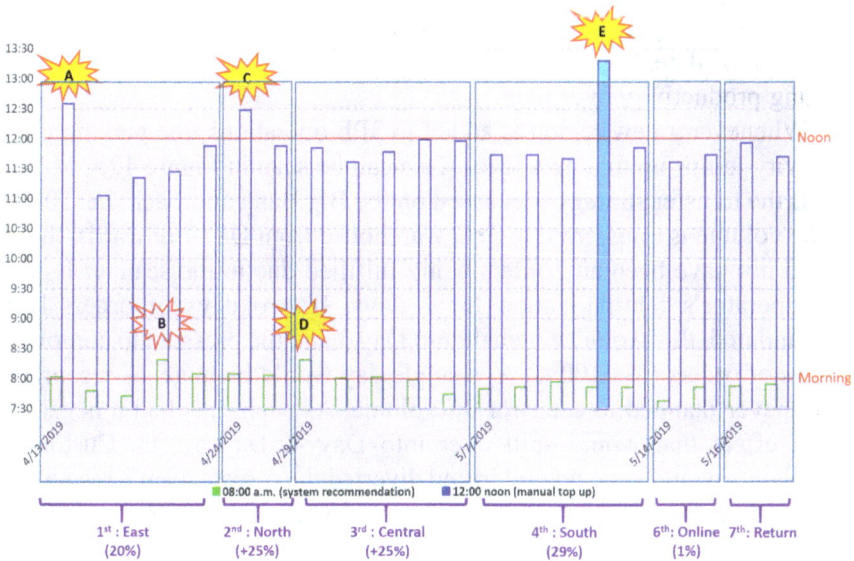

Figure 10. Impact of timing of orders received on business volume.

corresponding items in the sales order from the pick lines to be placed into the correct shipping cartons. Subsequently, there is a checker to check the accuracy of the pick by using a barcode scanner to scan every item in the shipping carton against the order in the computer. If all the items are correct, it then proceeds to a sealing process or else it is rejected and returned to the picker for re-pick.

To determine whether the picking functionality is properly transferred to the 3PL, the authors measured the picking productivity and accuracy. For picking productivity, the authors chose the maximum productivity in an hour of the day to be the yardstick because average picking productivity fluctuates according to the order quantity which varies daily. If the process know-how, resources, and tools are not properly transferred over, 3PL's productivity (max) and accuracy will be poor. During the first couple of days of go-live, the picking productivities were disastrous. On the first day, operations were intermittent. Workflow was frequently stopped due to data format issues between the new brand of scanners with 3PL's WMS and familiarization of the new operators on the operation procedures despite training. Zero orders were completed on Day 1. On the second day, although the scanner issue was resolved, the new crew of operators was still learning Company X's products and processes. Despite being slow, the crew managed to pick 104 pcs of items per man hour maximum on Day 2. And from Day 3 onwards, picking productivity continued to escalate. Figure 11 shows the morning and afternoon increase in picking productivity over time.

Whenever a new region is added to 3PL operations, the picking productivity jumps up to a new level. This can be seen in Figure 12.

If the transfer strategy was based on the Big Bang approach, i.e. 100% of the volume is transferred to 3PL Warehouse from Day 1 onwards, there would not have been any orders being fulfilled due to the scanner issues and operator's familiarization for 2 days. If two days of orders had accumulated, this would snowball into Day 3 and the pressure to complete 3 days of orders (3 × 100%) on a single day would be huge, as resources were never planned to cater for 3× volume. As such, this had a negative spiral effect that would spill over into Day 4, Day 5, etc. Customer complaints would have poured in and diverted the project team's resources to attend to the complaints, ultimately jeopardizing the entire transfer project.

5.12 *Operations: Packing completeness*

After picking, the cartons are sent for packing. At this packing station, all items in the carton are scanned and checked against the sales order in the system. If there is any wrong item or quantity detected, the whole carton

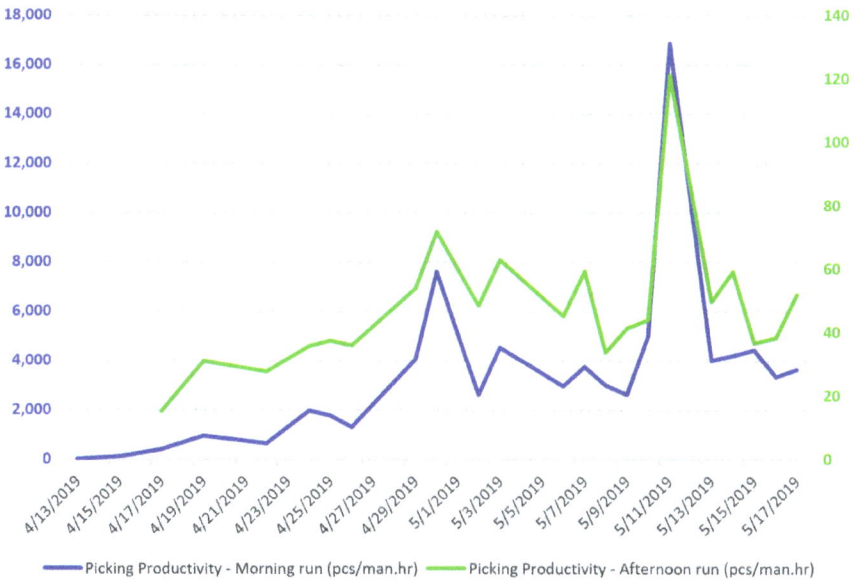

Figure 11. 3PL picking productivity.

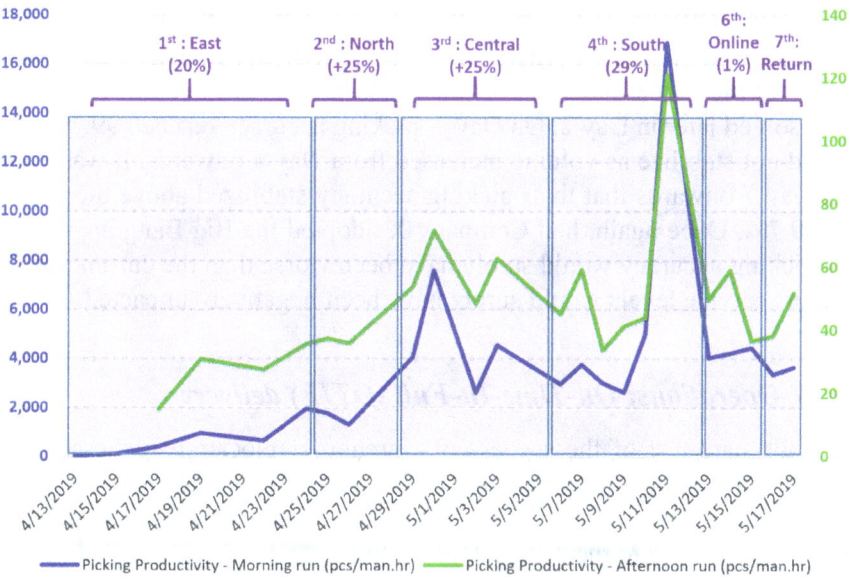

Figure 12. Impact of 3PL picking productivity on business volume.

Figure 13.   3PL picking accuracy.

is rejected and returned to the picking stations for re-pick. Only 100% accurate cartons may proceed and sent for sealing and subsequently sent to the shipping staging area to be stacked by delivery route for collection by the transporter. The picking accuracy on the first two days of go-live was disastrous as the 3PL new scanner's data format setting was not in sync with its WMS (Figure 13). Given the mismatched format, the system showed 0% accuracy because the scanner was sending one digit less (without the check digit) against the WMS record. Fortunately, this issue was solved later in Day 2. On Day 3, picking accuracy reached 99.7% but could not stabilize as volume increased from Day 4 onwards. It was only on Day 7 onwards that their picking accuracy stabilized above the target of 99.7%. Once again, had Company X adopted the Big Bang approach, its picking accuracy would surely have been worse than the current result and its service levels would surely have been negatively impacted.

5.13 *Operations: On-Time-In-Full (OTIF) delivery*

The ultimate test of the success of warehouse relocation is the service level of the destination warehouse. It is useless if the destination warehouse can perform its operations perfectly, but the orders cannot successfully be delivered to the end customers. Fortunately, Company X chose to use the same transporter. Thus, the changes are only limited to the new pick-up location and the handover procedure with the 3PL, instead of the Company X warehouse of origin. The transporter achieved 100% delivery OTIF level performance every day from the new destination warehouse, despite occasionally the 3PL warehouse could not get the orders ready to be loaded to the transporter by the agreed pick-up time. Figure 14 illustrates the delivery of OTIF results which speaks for itself.

This is instructive as it demonstrates that if there are fewer changes this translates into lower risks. If Company X had also changed the transporter in this warehouse relocation exercise, it is reasonable to expect worse delivery OTIF level results as compared to maintaining the same transporter.

5.14 *Recommendations*

In summary, the challenges faced during the Go-Live period included delays in the transmission of sales orders to the third-party logistics provider (3PL), zero picking on Day 1, low picking productivity in Phase 1 (with a gradual increase over time), sub-standard picking accuracy in the first 7 days, and delays in handing over packed orders to the transporter during high-volume days. The authors envision the possibility of these same issues arising during the three warehouse relocation approaches previously identified: (i) the Big Bang approach, (ii) gradual transfer by product category, and (iii) gradual transfer by region volume.

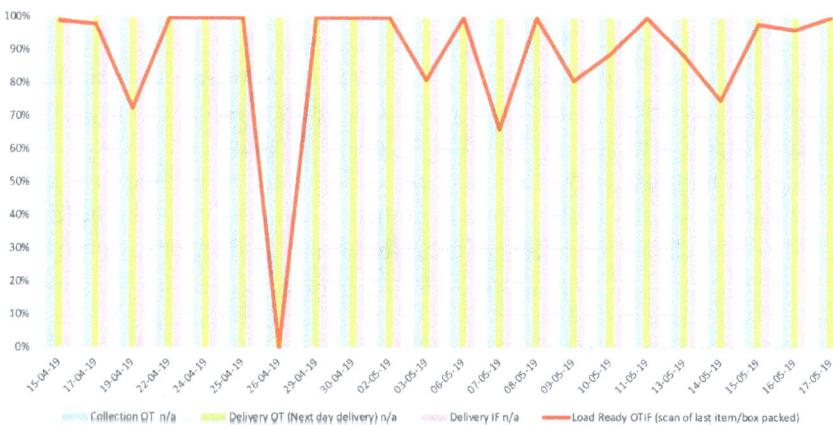

Figure 14. Delivery OTIF.

It is evident that if all five issues were to occur simultaneously on Day 1 of the Go-Live while using the Big Bang approach, it could prove disastrous for Company X. This could result in (i) zero sales for Company X on Day 1 as 100% of their sales orders could not be shipped from the new warehouse, (ii) a potentially low On-Time service level on 100% of sales orders in Phase 1 due to low picking productivity, leading to unprocessed orders spilling over into the next day, and (iii) a potentially low In-Full service level on 100% of sales orders within the first 7 days due to sub-standard picking accuracy, potentially resulting in stock losses.

Hence, the two gradual transfer approaches are superior in this case compared to the Big Bang approach. The gradual transfer approach enables the spreading of these issues to a smaller volume of orders affected during the warehouse relocation, reducing potential losses, such as a 20% loss of total sales orders during Phase 1. Finally, it is suggested that the gradual transfer approach by region volume is the superior choice. This approach increases the complexity of the processes for sales orders that need to be fulfilled from two warehouses, potentially negatively impacting efficiency.

5.15 *Managerial implication*

Company X's decision to relocate its warehouse and outsource warehousing activities was part of a broader strategy aimed at optimizing its assets and streamlining its workforce to enhance overall performance. This move allowed Company X to become more efficient and agile and focus on its core competency, which is marketing and selling products through its MLM network. However, the implementation of warehouse relocation projects in such circumstances carries substantial risks that can negatively impact responsiveness and service levels (Nagahan and Akın, 2018). Previous warehouse relocation projects at Company X resulted in significant financial losses, highlighting the inherent complexity and risks associated with such endeavors. To address these risks and ensure a successful relocation, a resilient warehouse relocation strategy is crucial. The potential for lost sales and disruptions to service levels necessitates careful planning (Petersen and Aase, 2016).

The project addressed challenges, including a tight timeline for completion, the need for redundancy notices to warehouse operators, and the requirement to maintain order fulfillment service levels. A thorough analysis of Company X's existing operations was conducted to identify potential risks. Change factors at each stage of the relocation were evaluated, and a comprehensive risk identification process was undertaken. Each identified risk was assessed in terms of probability, and corresponding risk mitigation action plans were developed. Several relocation approaches were considered, and a gradual transfer of operations by region was chosen due to its lower operational risks. The success of the warehouse relocation project hinged on the performance of operations at the new warehouse (Petersen and Aase, 2016). KPIs were used to assess warehousing operations post-relocation. Initial challenges were observed in the early days of the relocation, affecting picking productivity, packing completeness, and OTIF (On-Time In-Full) delivery measures. However, these issues were swiftly addressed, and operations improved from the third day onward.

5.16 *Limitations and future research*

The experiences during the initial teething period underscore the importance of resilience in warehouse relocation strategies. Despite careful analysis, risk mitigation planning, and method selection, unforeseen issues can arise, impacting the relocation process. This highlights the critical need for a resilient approach when embarking on high-risk and complex projects like warehouse relocation to minimize their impact on overall company operations. While Company X provided compensation packages and absorbed some workers into new 3PL operations, a comprehensive change management plan for redundant workers was lacking due to budget constraints. Future research in warehouse relocation projects should consider developing a robust change management plan to address the risks associated with redundant workers and further enhance the resilience of warehouse relocations.

References

Ageron, B., Bentahar, O., and Gunasekaran, A. (2020). Digital supply chain: Challenges and future directions. *Supply Chain Forum: An International Journal*, 21(3), 133–138.

González-Hernández, Martínez-Flores, Sánchez-Partida, Gibaja-Romero. (2019). Relocation of the distribution center of a motor oil producer reducing its storage capacity: A case study. *Simulation*, 95(11), 1097–1112.

Gurtu, A. and Johny, J. (2021). Supply chain risk management: Literature review. *Risks*, 9(1), 16.

Kotonen, T. (2017). *Managing Operational Risks During Warehouse Relocation Project*, MBA thesis.

Kuster, J., Bachmann, C., Hubmann, M., Lippmann, R., and Schneider, P. (2023). *Project Management Handbook: Agile-Traditional-Hybrid* (Vol. 2). Springer.

Nagahan Yaylali-Salcıoğlu and Akın Marşap (2018). The impact of warehouse relocation on the supply chain department. *International Journal of Business Marketing and Management*, 3(8), 30–45.

Petersen, C.G. and Aase, G.R. (2016). Issues in Distribution Center Relocation. *Open Journal of Business and Management*, 4, 7–13.

Singh, N. (2022). Developing business risk resilience through risk management infrastructure: The moderating role of big data analytics. *Information Systems Management*, 39(1), 34–52.

Zapata, C., Garnica, A., Mota, R., Alvarez, F., Flores, L., Partida, D. (2020). Warehouse relocation of a company in the automotive industry using P-median. *Advances in Science, Technology and Engineering Systems Journal*, 5(3), 576–582.

Chapter 10

Warehouse Performance Management and Continuous Improvement

Learning Outcome

By the end of this topic, you should be able to do the following:

1. Measure warehouse performance.
2. Identify people, process, and procedure.
3. Build a culture of continuous improvement.

1. Introduction

From the user-owned perspective, a warehouse is viewed as a cost function. In contrast, a third-party logistics (3PL) perspective treats it as a revenue function. Nevertheless, in both scenarios, operating a warehouse incurs running costs that involve the coordinated use of human resources and machinery. Therefore, a warehouse must remain competitive by effectively managing **productivity, quality, and cycle time performance**, as these factors directly influence the company's bottom line. As the saying goes, "What gets measured gets managed." By measuring performance, we can determine whether we are improving internally or gaining a competitive edge.

There are two types of measurements: **quantitative and qualitative**. In this context, we focus on quantitative measurements — those that can be objectively assessed. For example, a 10% increase in inventory accuracy provides a clear metric for evaluation, as opposed to simply categorizing performance as good or bad. Before we explore key performance indicators (KPIs), it is good to understand the key concepts of **Efficiency and Effectiveness**.

Efficiency is fundamentally about executing tasks correctly. i.e. **doing things right**. It involves the capability to complete a job while utilizing the fewest resources possible, including time, money, and effort. An efficient process aims to maximize output while minimizing input. Consider two pickers: Picker A and Picker B. Picker A picks 100 items per hour, whereas Picker B picks 110 items in the same timeframe. This means Picker B is 10% more efficient or productive than Picker A, achieving greater results without increasing the time spent.

Effectiveness, on the other hand, emphasizes **doing the right things**. It evaluates how well a goal or objective is accomplished, regardless of the resources used. An effective process successfully achieves its intended outcomes. For example, in a warehouse environment, if the team recognizes that certain products are frequently ordered together, placing these items near each other can streamline the picking process and reduce fulfillment times by 10%. However, this improvement may require the team to invest additional time in analyzing order patterns and reorganizing existing inventory to consolidate these items. While this upfront effort may take longer, it ultimately leads to significant long-term benefits in operational efficiency and customer satisfaction.

In essence, **efficiency and effectiveness complement each other**. Efficiency focuses on optimizing resources and streamlining processes, whereas effectiveness is centered on achieving meaningful outcomes that align with strategic goals. Striking a balance between the two can lead to optimal performance across various activities. Ideally, one should prioritize effectiveness first, followed by efficiency. When evaluating warehouse performance, we can utilize efficiency or productivity metrics. This assessment begins with the receiving operations and extends to the final dispatch of completed, picked, and packed shipments or orders. At each stage of this process, we can measure the performance of various operations and activities. This comprehensive approach allows us to assess the concept of the perfect order in our warehouse operations. We delve deeper into the perfect order concept after discussing the Key Performance Indicators (KPIs) for each operation or activity.

The primary operation of a warehouse involves receiving shipments or inbound from suppliers or transferring goods from another owned warehouse. To effectively evaluate this process, we focus on three key metrics: **accuracy, speed, and cost** of receiving shipments. These can be quantified by measuring each shipment, line item, or unit (each carton/pallet). Selecting an appropriate Unit of Measurement (UOM) is crucial for meaningful KPI assessment. If we select shipment, it will be a high-level measurement as each shipment consists of multiple line items with various quantities. If we opt for line items, we encounter challenges since some line items may consist of varying quantities. Additionally, a company might manage hundreds or even thousands of line items in its inventory. While we could calculate averages for each line item to establish a uniform quantity, this approach has limitations. The final average may differ for each shipment, leading to inconsistencies in measurement. Given these complexities, measuring by individual units is generally a more effective approach than using line items. However, it's important to note that this method may still present challenges; different sizes of units can result in varying handling times. Thus, while unit measurement offers a clearer picture, it's essential to consider the context of the units being handled.

2. Measuring Warehouse Performance

To evaluate warehouse performance, metrics should cover the entire operational flow — from **receiving** shipments to the final **dispatch** of orders. Performance can be assessed using **efficiency metrics** and the concept of the **perfect order**, which we detail later.

Key Performance Indicators for Inbound Operations
Inbound operations include **receiving** and **put-away** processes. To evaluate these operations, KPIs focus on **accuracy**, **speed**, and **cost**. Selecting the appropriate **Unit of Measurement (UOM)** (e.g. shipment, line item, or unit) is critical for meaningful assessment.

2.1 *Inbound metrics* (*receiving and put-away*)

Inbound metrics focus on the accuracy, speed, and cost of bringing goods into the warehouse and storing them efficiently.

1. **Receiving Accuracy**
    Definition: Receiving accuracy measures how accurately goods are received compared to the purchase order or shipping documentation.
    Purpose: It ensures that the correct quantity and items have been received. This reduces downstream errors, such as stock discrepancies or incorrect inventory levels.
    Example Application: If 1,000 units were received but 10 were damaged or miscounted, receiving accuracy would be
    Accuracy = 990/1000 = 99%

2. **Put-away Accuracy**
    Definition: Put-away accuracy evaluates how accurately items are stored in their assigned locations after receiving.
    Purpose: It reduces picking errors by ensuring that items are placed in the correct locations and match the warehouse management system (WMS) records.
    Example Application: If 950 units out of 1,000 were put in the correct location, put-away accuracy would be
    Accuracy = 950/1000 = 95%

3. **Receiving On-Time**
    Definition: Receiving on time tracks the timeliness of the receiving process based on predefined time standards.
    Purpose: It minimizes unloading delays, which could disrupt downstream operations like put-away or order picking.
    Example Application: If 90 out of 100 expected shipments were received within the target time, on-time receiving performance is
    On-time receiving = 90/100 = 90%

4. **Receipts Per Man-Hour**
    Definition: Receipts per man-hour measures the number of items or units received per hour of labor.
    Purpose: It indicates the efficiency of the receiving team, identifying opportunities to streamline labor-intensive processes.
    Example Application: If 500 units were received by a team working 10 hours, productivity is
    Receipts per Man-hour = 500/10 = 50 (units per hour)

5. **Time Taken to Receive Each UOM**
    Definition: Time taken to receive each UOM calculates the time and associated labor cost to receive a single unit of measure (UOM).
    Purpose: It helps evaluate cost-efficiency and identify processes that need improvement.

Example Application: If receiving 1,000 units takes 5 hours at $20/ hour,

Cost per UOM = (5 × 20)/1000 = $ 0.10 per unit

2.2 *Outbound metrics (picking, packing, and shipping)*

Outbound metrics focus on fulfilling customer orders efficiently and accurately.

1. **Picking Accuracy**

   Definition: Receiving accuracy measures how many items are picked correctly from storage compared to customer orders.

   Purpose: It prevents order errors that can result in customer dissatisfaction or returns.

   Example Application: If 970 items out of 1,000 were picked correctly, the accuracy would be

   Accuracy – 970/1000 = 97%

2. **Packing Accuracy**

   Definition: Packing accuracy tracks whether items are packed correctly according to customer specifications (e.g. right products, correct quantities, and proper packaging).

   Purpose: It reduces shipping errors, damage, and customer complaints.

   Example Application: If 98 out of 100 orders were packed correctly, packing accuracy would be

   Accuracy = 98/100 = 98%

3. **Shipping Accuracy**

   Definition: Shipping accuracy evaluates the accuracy of shipments dispatched to customers, ensuring they meet the order specifications.

   Purpose: It ensures complete and correct delivery, preventing returns and improving customer satisfaction.

   Example Application: If 990 out of 1,000 shipments were accurate, shipping accuracy is

   Accuracy = 990/1000 = 99%

4. **Picked, Packed, and Shipped On-Time**

   Definition: Picked, packed, and shipped on-time tracks whether each step in outbound operations meets the expected timeline.

   Purpose: It ensures orders are fulfilled promptly to meet customer expectations.

Example Application: If 85 out of 100 orders were picked, packed, and shipped on time,
On-Time Rate = 85/100 = 85%

5. **Outbound Per Man-Hour**
   Definition: Outbound per man-hour measures the number of outbound units processed per hour of labor for each activity (picking, packing, and shipping).
   Purpose: It highlights areas where efficiency can be improved.
   Example Application: If 300 orders were packed in 5 hours,
   Packed per Man-Hour = 300/5 = 60 orders per hour

6. **Cost per UOM for Outbound Activities**
   Definition: Cost per UOM for outbound activities calculates the labor cost for picking, packing, or shipping one unit.
   Purpose: It assists in analyzing and managing operational expenses.
   Example Application: If packing 1,000 orders takes 10 hours at $25/hour,
   Cost per UOM = (10 × 25)/1000 = $0.25 per order

2.3 *Inventory management metrics*

Inventory metrics assess the accuracy and reliability of stock records, crucial for ensuring smooth operations.

1. **SKU Quantity Accuracy**
   Definition: SKU quantity accuracy measures how accurately the physical inventory matches recorded inventory levels for each SKU.
   Purpose: It prevents stockouts or overstocking by ensuring reliable inventory data.
   Example Application: If 90 SKUs out of 100 match the records,
   SKU Quantity Accuracy = 90/100 = 90%

2. **Location Accuracy**
   Definition: Location accuracy racks how accurately items are stored in the correct locations according to the system records.
   Purpose: It reduces errors and delays during picking and put-away processes.
   Example Application: If 48 of 50 SKUs are in the correct location,
   Location Accuracy = 48/50 = 96%

2.4 *The perfect order metric*

The **perfect order metric** combines all critical operational metrics into a single composite score to evaluate overall warehouse performance. It accounts for receiving, put-away, picking, packing, and shipping. Each process's performance contributes to the final metric. A high perfect order score indicates an end-to-end efficient and effective operation, while a low score signals areas for improvement.

Example:
If each operation has the following accuracies:

- receiving: 99.5%
- put-away: 99%
- picking: 98%
- packing: 99.5%
- shipping: 99.5%

the perfect order score is 99.5% × 99.0% × 98.0% × 99.5% × 99.5% = 95.57%.

These metrics provide warehouse managers with critical insights to pinpoint inefficiencies, benchmark performance, and drive continuous improvement initiatives. For example, when analyzing the perfect order computation, we discovered that our accuracy was substantially lower than anticipated when evaluating the scores of individual activities. This finding underscores the effectiveness of the perfect order metric as a holistic approach to assessing overall warehouse performance, which directly influences customer satisfaction. Similarly, we applied the same calculation method to measure on-time delivery (speed), enabling us to determine the percentage of perfect orders achieved within the required timeframes.

3. People, Process, and Procedure

The continuous improvement methodology originates from the Japanese concept of Kaizen, which emphasizes making consistent, incremental improvements over time, whether weekly, monthly, or annually. For example, dedicating just 30 minutes to an hour each day to studying warehouse management can result in substantial knowledge growth over a year. This disciplined approach can lead to a deep understanding of effective practices, aligning with lean management principles that prioritize eliminating waste and optimizing processes by removing non-value-added activities. Before implementing continuous improvement initiatives, it is crucial to establish a solid foundation using the 3P Framework: Process, Policy, and Procedure. This framework ensures operational efficiency, safety, and consistency in warehouse management.

3.1 *Process*

The "Process" component refers to the sequence of steps involved in warehouse operations, such as receiving, storing, picking, packing, and shipping goods. Clearly defining these processes is vital for ensuring smooth and efficient workflows. For instance, the receiving process might involve verifying incoming shipments against purchase orders, inspecting for damages, and updating inventory systems. Mapping out these steps allows organizations to identify inefficiencies, streamline workflows, and ensure employees clearly understand their roles.

3.2 *Policy*

Policies are the overarching rules and guidelines that govern warehouse operations. They establish behavioral and performance standards for team members, encompassing areas such as safety protocols, inventory management, and workplace conduct. Well-documented policies not only reduce errors and improve compliance but also foster a safer, more organized environment.

3.3 *Procedure*

Procedures provide detailed, step-by-step instructions for completing specific tasks within a process. Formalized as Standard Operating Procedures (SOPs), these documents guide activities such as inventory counting and order fulfillment. SOPs ensure consistency across operations, and their regular review and updates help organizations stay aligned with evolving technologies and regulatory requirements.

By integrating these three components — Process, Policy, and Procedure — warehouses can enhance efficiency, minimize errors, and significantly boost productivity. This structured approach ensures that continuous improvement efforts are built on a strong, reliable foundation.

3.4 *Transforming a chaotic warehouse environment*

In my previous role managing warehouse operations within a 3PL environment, we utilized the 3P framework to revamp a disorganized Regional Distribution Center (RDC) for a personal hygiene company. This RDC managed 26,000 pallet locations, operated 24/6, and employed 100 warehouse staff along with 10 administrative personnel working 12-hour shifts.

On my first day, the facility was in chaos. Debris, papers, and plastic wraps were scattered across the warehouse, and Material Handling Equipment (MHE) such as forklifts and reach trucks was moved about without defined pedestrian walkways. Despite the team appearing busy, there was no structured approach to operations, and safety was a low priority. Workers picked SKUs based on convenience rather than adhering to the Warehouse Management System (WMS), leading to disorganized inventory and an accuracy rate of only 95%, significantly below the 99.5% stipulated in our contract.

To address these issues, we implemented the Japanese 5S Housekeeping Technique: Sort, Set, Shine, Standardize, and Sustain. We created training materials and conducted workshops for all 110 staff members to familiarize them with the methodology. Following classroom sessions, we organized a "Waste Walk" to identify inefficiencies in both the warehouse and office areas. Initially, we held daily stand-up meetings to address immediate operational issues, such as incorrect item picking or misplaced inventory. Once the team became proficient with the correct processes, we reduced meetings to a weekly schedule, allowing for more efficient communication and sustained improvement.

3.5 *Enhancing safety and employee engagement*

Beyond process improvements, we introduced a comprehensive warehouse safety program. This initiative included training on safe work practices to prevent lifting injuries, restricting MHE operations to trained employees, and conducting fire safety drills to familiarize staff with escape routes and emergency procedures. To reinforce safety practices, each shift began with a three-minute safety briefing during daily toolbox meetings.

We also established accountability policies to address issues such as neglecting work areas, lacking a safety mindset, and mishandling inventory. Alongside corrective measures, we implemented positive reinforcement strategies to reward exemplary behavior. Employees who reported hazards or demonstrated strong safety practices were recognized with small monetary awards and gifts during toolbox briefings. Additionally, we launched a quarterly competition between AM and PM shifts to determine the best-performing team. Winning teams were celebrated with a buffet dinner, a trophy, and shared monetary prizes, fostering a sense of healthy competition and teamwork.

3.6 *Updating procedures and ensuring ISO compliance*

With processes and policies firmly in place, we focused on updating outdated SOPs to reflect current practices. As an ISO 9000-certified company, we adhered to the principle of "say what you do and do what you say," undergoing annual or surveillance audits to maintain compliance. From these SOPs, we developed detailed Work Instructions (WIs), which outlined specific tasks, roles, and responsibilities step-by-step. To enhance clarity, we adopted a swim-lane format to visually map the process flow, associated documents, timelines, and individual responsibilities. These updated documents became the foundation for training both new and existing employees, ensuring consistent alignment across the workforce.

3.7 *Optimizing processes with the flow process chart*

Once the 3P framework and 5S methodology were firmly established, we introduced the flow process chart to optimize efficiency across inbound and outbound activities. Inbound processes included receiving, staging, and putting away stock, while outbound activities encompassed order processing, picking, packing, staging, and shipping.

To create the flow process chart, we began by mapping the current processes using standard symbols to represent operations, transport, delays, inspections, and storage activities. We calculated the total time and distance for all tasks, identifying delays and transport as "waste" within the lean management framework. By eliminating or reducing these wasteful activities, we repositioned value-added tasks — such as operations, inspections, and storage — closer together, streamlining the overall workflow (see Figure 1).

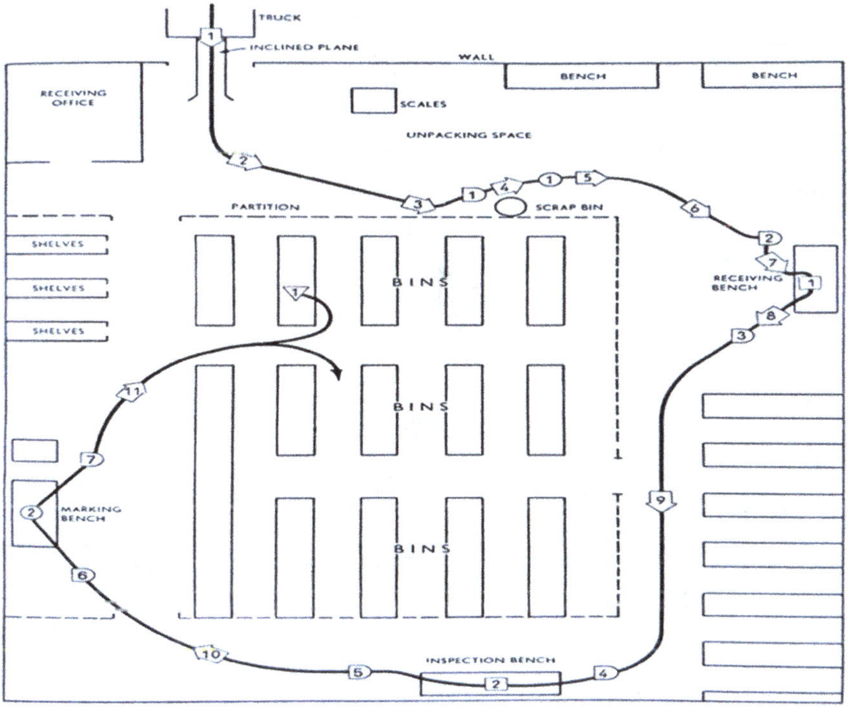

Figure 1. Flow Diagram (Present Method).

Source: Adapted from Introduction to work study (3rd ed.), by International Labour Office (1979, p. 113).

Figure 2 illustrates that there were 11 units of transport and 7 units of delays, resulting in a total distance of 56.2 meters and a time expenditure of 174 minutes to complete these tasks. By reorganizing the layout and activities within the warehouse and reassessing the overall effectiveness of the processes, we streamlined operations, reducing the number of operational processes to 6 units, transport processes to 2 units, and inspections to just 1 unit. The reduction in inspections to a single unit was possible because we initially implemented numerous checks and balances; however, once the processes stabilized, these were no longer necessary. In a previous role at a manufacturing firm producing AT&T cordless phones for home use, we encountered a similar situation. As a startup, the test and tune engineer had introduced many redundant test scripts at each assembly station. Once the processes were stabilized, we successfully eliminated

FLOW PROCESS CHART			MAN/MATERIAL/EQUIPMENT TYPE			
CHART No. 3 SHEET No. 1 OF 1			S U M M A R Y			
Subject charted: Case of BX 487 Tee-pieces (10 per case in cartons)			ACTIVITY	PRESENT	PROPOSED	SAVING
			OPERATION O	2		
ACTIVITY: Receive, check, inspect and number tee-pieces and store in case			TRANSPORT ⇒	11		
			DELAY D	7		
METHOD: PRESENT/PROPOSED			INSPECTION □	2		
LOCATION: Receiving Dept.			STORAGE ▽	1		
OPERATIVE(S): CLOCK No.			DISTANCE (m)	56.2		
See Remarks column			TIME (man-h.)	1.96		
CHARTED BY: DATE:			COST LABOUR	$10.19		
APPROVED BY: DATE:			MATERIAL	—		
			TOTAL	$10.19		

DESCRIPTION	QTY. 1 case	DIST-ANCE (m)	TIME (min)	O	⇒	D	□	▽	REMARKS
Lifted from truck: placed on inclined plane		1.2							2 labourers
Slid on inclined plane		6	10						2 „
Slid to storage and stacked		6							2 „
Await unpacking		—	30						
Case unstacked		—							
Lid removed: delivery note taken out		—	5						2 „
Placed on hand truck		1							
Trucked to reception bench		9	5						2 „
Await discharge from truck		—	10						
Case placed on bench		1	2						2 „
Cartons taken from case: opened: checked replaced contents		—	15						Storekeeper
Case loaded on hand truck		1	2						2 labourers
Delay awaiting transport		—	5						
Trucked to inspection bench		16.5	10						1 labourer
Await inspection		—	10						Case on truck
Tee-pieces removed from case and cartons: inspected to drawing: replaced		1	20						Inspector
Await transport labourer		—	5						Case on truck
Trucked to numbering bench		9	5						1 labourer
Await numbering		—	15						Case on truck
Tee-pieces withdrawn from case and cartons: numbered on bench and replaced		—	15						Stores labourer
Await transport labourer		—	5						Case on truck
Transported to distribution point		4.5	5						1 labourer
Stored									
TOTAL		56.2	174	2	11	7	2	1	

Figure 2.   Receiving activities using Flow Process Chart (Present Method).

Source: Adapted from Introduction to work study (3rd ed.), by International Labour Office (1979, p. 114).

30% of the total time from the assembly line. This optimization was crucial as we needed to scale production from 5 units to 1,000 per day, ultimately saving both time and effort. After the improvement processes were implemented, as shown in Figures 3 and 4, there was a significant improvement in the transportation of 45.5%, delays of 71.4%, and 50% of inspection activities which were eliminated.

Once we have enhanced the overall flow of inbound operations, we will apply the same techniques to outbound operations. You might be surprised by the number of non-value-added activities present in the

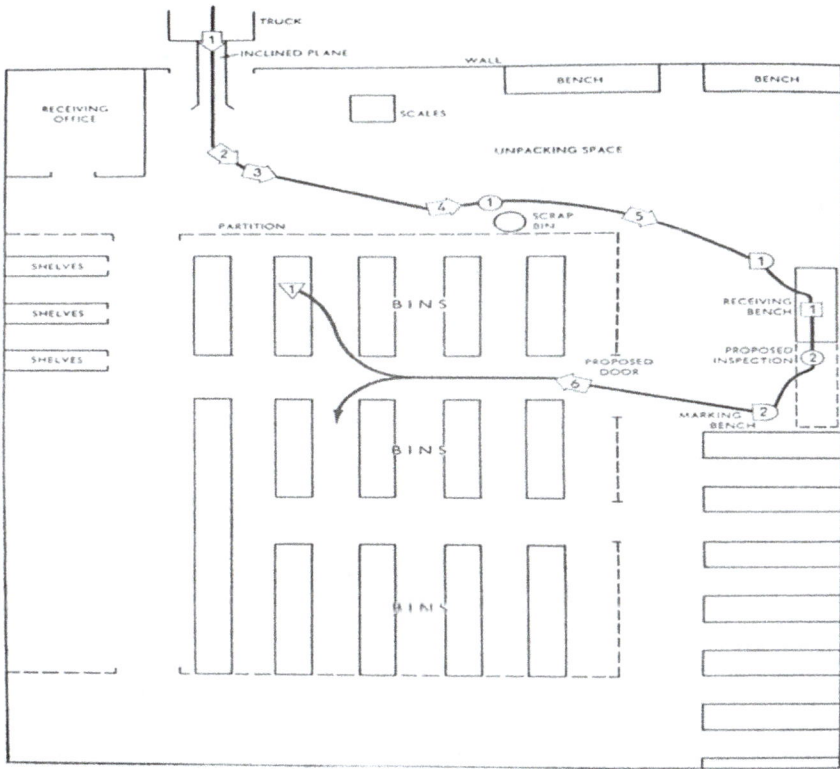

Figure 3. Receiving activities using Flow Process Chart (Proposed Method).

Source: Adapted from Introduction to work study (3rd ed.), by International Labour Office (1979, p. 116).

existing processes. Whenever there are changes in the overall workflow, it is advisable to create a revised flow process chart to streamline these processes effectively. In fact, regularly reviewing all process flows is a best practice for continuously improving overall efficiency. After making improvements to the "big picture" of warehouse operations, it's time to focus on individual activities within each process. For instance, we should examine the receiving activities that occur when a truck or container arrives at the security checkpoint, as well as the unloading procedures in the loading bays.

There are a variety of tools and techniques available to improve operational processes. Some of the most effective include Kaizen,

FLOW PROCESS CHART				MAN/MATERIAL/~~EQUIPMENT~~ TYPE				
CHART No. 4　　SHEET No. 1　　　OF 1					S U M M A R Y			
Subject charted:				ACTIVITY		PRESENT	PROPOSED	SAVING
Case of BX 487 tee-pieces (10 per case in cartons)				OPERATION ◯		2	2	—
				TRANSPORT ⇨		11	6	5
ACTIVITY: Receive, check, inspect and number				DELAY 𝔻		7	2	5
tee-pieces store in case				INSPECTION ☐		2	1	1
METHOD ~~PRESENT~~/PROPOSED				STORAGE ▽				
				DISTANCE (m)		56.2	32.2	24
LOCATION: Receiving Dept.				TIME (man-h.)		1.96	1.16	0.80
OPERATIVE(S)　　　　　CLOCK No				COST per case				
See Remarks column				LABOUR		$10.19	$6.03	$4.16
CHARTED BY　　　　　DATE:				MATERIAL		—	—	—
APPROVED BY　　　　DATE				TOTAL		$10.19	$6.03	$4.16

DESCRIPTION	QTY	DIST-ANCE	TIME	SYMBOL					REMARKS
	1 case	(m)	(min)	◯	⇨	𝔻	☐	▽	
Crate lifted from truck; placed on inclined plane		1.2							2 labourers
Slid on inclined plane		6	5						2　˝
Placed on hand truck		1							2　˝
Trucked to unpacking space		6	5						1 labourer
Lid taken off case		—	5						1　˝
Trucked to receiving bench		9	5						1　˝
Await unloading		—	5						
Cartons taken from case: opened and tee-pieces placed on bench; counted and inspected to drawing		—	20						Inspector
Numbered and replaced in case		—							Stores labourer
Await transport labourer		—	5						
Trucked to distribution point		9	5						1 labourer
Stored		—	—						
TOTAL		32.2	55	2	6	2	1	1	

Figure 4.　Flow Diagram (Proposed Method).

Source: Adapted from *Introduction to work study* (3rd ed.), by International Labour Office (1979, p. 117).

DMAIC, the PDCA cycle, and cause–effect analysis. Among these, we particularly favor the Kaizen approach, the 5W2H brainstorming technique, and the use of the weekly Kaizen Minutes of Meeting (MoM). Additionally, with any improvement project, issues or problems need to be addressed, and we recommend using the fishbone diagram or cause–effect analysis method to analyze five environments or milieu (man, machine, material, method, and measurement) existing in a warehouse environment. This series of steps helps clarify thought processes by documenting ideas and facilitating weekly follow-up meetings. This structured approach ensures that we stay on track to achieve specific improvement goals in a timely manner.

As with any improvement initiative, it is essential for management to be enthusiastic about launching the program. The program lead is responsible for developing comprehensive plans that aim to gain support from employees, especially those on the ground. For example, we were once assigned to lead a project focused on reducing company-wide overtime by 10%, and the resulted in yearly savings of $200,000.0 This initiative had a direct impact on the take-home pay of our ground staff, leading to initial resistance from the team. However, by discussing the importance of cost competitiveness and efficiency within the 3PL environment, we were able to secure their support.

Let's begin with the first step by using the Kaizen form with a 10% overtime deduction as an example (Figure 5). The sample Kaizen form in the following outlines the steps taken to describe the current situation and the issues or obstacles encountered in achieving the proposed condition.

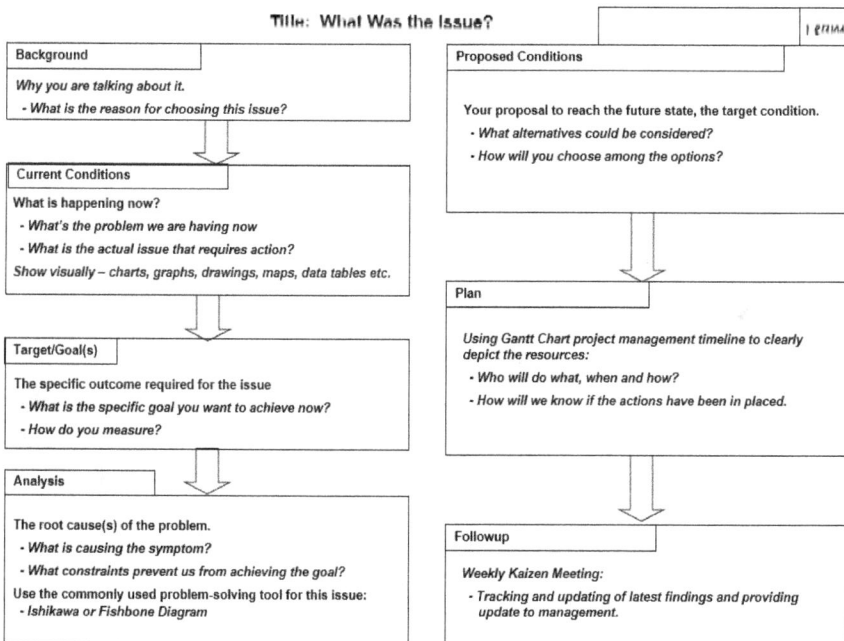

Figure 5. Kaizen form template.

Source: Adapted from A3 Report Template from a Case Study conducted for one of the SME (Small Medium Enterprise).

Using the same format, we translated our outcome in the title but to fill up the form meaningfully, we need support from the project team from the warehouse operations staff—warehouse manager to the warehouse assistants running the ground operations. Prior to filling up the form, we need to set up a project team consisting of the warehouse manager, executive, order administration, and warehouse assistant guided by the project lead. The project lead is the go-to person for guiding the team in using process improvement tools. In some companies, the operations excellence team will guide the team in managing the improvement project.

After filling in the title, it is time to use the 5W2H brainstorming technique to sift out all the issues with regard to the overtime. This customer carries premium mattresses and its volumes increased 30% year on year. With the increase in volume, the warehouse space, especially the incoming staging area, and with existing manpower, we would not be able to handle this during the working hours from 08.30am to 05.30pm: 8.0 hours of effective working hours with one-hour breaks. Using the 5W2H brainstorming technique, we discovered as per Figure 6.

Above are two brainstorming sessions which we had gone through, and typically, it took 2 hours for each session. Once the brainstorming ideas had been put in place in writing, we would deploy the fishbone diagram or cause–effect diagram analysis to further analyze the causes and effects which contributed to the increase in overtime cost. Using this technique allows us to drill down the cause of the problem in 5 M (milieu) as mentioned earlier. A sample cause–effect analysis format is as per Figure 7.

After identifying each effect that influenced the cause, we develop a plan to address each cause impacting the issue at hand. For instance, we outline a specific cause that affects the effect. In addition to identifying the root cause or issue, we also highlight the symptoms along with proposed solutions, ensuring that each resolution has a clear timeline for completion (Figure 8).

From our overtime reduction case study, the cause–effect analysis and cause–effect solution are as per Figures 9 and 10.

Before completing the final Kaizen form, we would need to record all the meetings and plan of actions in a weekly Kaizen Minute of Meeting (MoM). See the snippets of the OT deduction MoM (Figure 11).

The project mentioned above took approximately two months to complete, during which we held weekly meetings. Minutes from these meetings were documented and circulated to all relevant parties to ensure close monitoring of outstanding tasks. At this point, we have only eight weeks remaining to finish the project. Not all factors contributed equally to the

Reduce OT hours at 10% – Customer A

5W2H Brainstorming:

1. **What** are the major constraints incurred high overtime?
   A. More than 1 container or Malaysia (MY) truck incoming in one day as one container will take 4 hours to complete the processes of unstuff, sort, QC, labeling and put-away.
   B. Customer confirmed order late after 3.00pm resulted late picking activities.
   C. Local distribution loading delayed the incoming as driver came late in the morning.

2. **Why** are there with major constraints?
   A. Inbound receiving start at 10am or later after releasing daily local distribution.
   B. Sorting SKUs to segregate on each pallets base on the packing list against physical supplier label.
   C. Slow progress due to items are too heavy, especially bed frame.
   D. Completion of sorting & receiving can be 3 hours or more for 40ft container or truck.
   E. Daily order picking delay by Inbound receiving.

3. **Where** are the major constraints?
   A. Not enough storage locations.
   B. Insufficient staging area.
   C. Printing of product labeling and label on each piece.
   D. Quality check after receiving.
   E. Put-away after quality check.
   F. Shortage of manpower resources.

4. **Who** was affected?
   A. Warehouse
   B. Customer

5. **When** did it happen?
   A. Volume increase in early 2021.

6. **How** did it happen?
   A. Customer expand business model (Hotel) with additional 30% more pallets locations.
   B. Volume surge up (43% increase on Inbound/outbound 2020 Vs 2021).
   C. 50% more product inspection activities by Customer's Products and Marketing personnel.

7. **How** much impact did it cost on Overtime?
   A. We spent $10,000 (2021) Vs $8,000 (2020), increased of 25%
   B. Additional manpower support after 05.30pm
   C. 2 hours OT man-hours per day for 4 persons

8. **How** can the situation be corrected and to reduce overtime?
   A. Customer to control/schedule on their monthly shipment in order not to crash on the same date.
   B. Inbound TAT tally sheet reporting to Customer at N+3 days, propose plan:
      (i) N+0 day – Unstuff/unloading to pallets by SKUs level and product labeling
      (ii) N+1 day – Quality check and put-away
      (iii) N+2 days – Submit Tally Sheet to Customer
      (iv) Proposed to have one more day based on MY container late arrival after 1.00pm.

   C. Deploy RF scanning to speed up inbound and outbound – customer is exploring.
   D. Factory fitted with barcode labels on the incoming products – customer is exploring.
   E. Overflow extra storage volume to second floor

Figure 6. 5W2H Brainstorming for 10% OT deduction.

Note. You can use more that 5W and 2H open ended questioning if neccessary.

Source: Adapted from a Case Study conducted for one of the SME (Small Medium Enterprise).

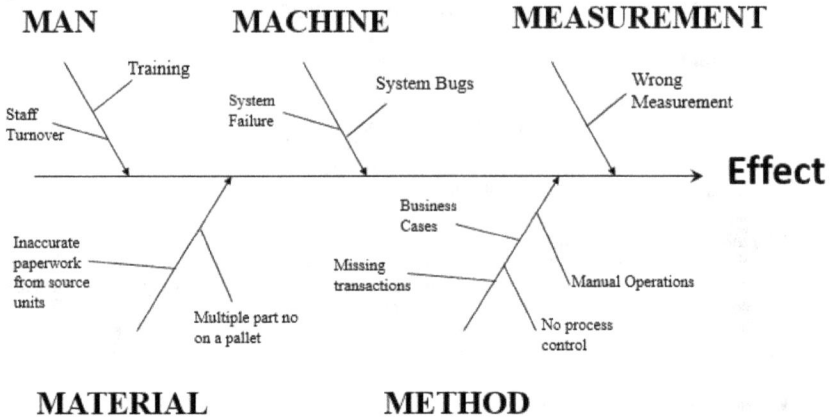

Figure 7.   Sample cause–effect diagram analysis.

Note: This is also known as Fishbone Diagram. The Fishbone represents the Causes and the head is the Effect.

Source: Adapted from a Case Study conducted for one of the SME (Small Medium Enterprise).

- **Root Cause #1**
  - Complex Business Cases

- **Symptoms**
  - Manual Entries
  - Missing transactions
  - Data does not tally

- **Solution**
  - System Modification
  - Intermediate fix
    - Invoice no

- **Timeline**
  - 2 weeks for system modification

Figure 8.   Sample cause–effect solution.

Source: Adapted from a Case Study conducted for one of the SME (Small Medium Enterprise).

10% deduction in overtime (OT). Therefore, we employ the Pareto principle analysis, or the 80/20 rule, which suggests that a small percentage of causes often leads to a large percentage of effects. Specifically, we examine the 20% of OT types that account for 80% of OT hours. While the exact ratio may not be precisely 20%, the concept remains valid.

In this case study, 50% (3 out of 6 order types) of the OT types were responsible for 80.4% of the total OT amounts. To ensure timely completion, we must concentrate on achieving the most significant impact. As one of my professors used to say, "It is better to have an imperfect answer on time than a perfect answer too late!". Once we had identified the three significant impacts, we would put in the priority to achieving the maximum outcome as per the Pareto analysis (Figure 12).

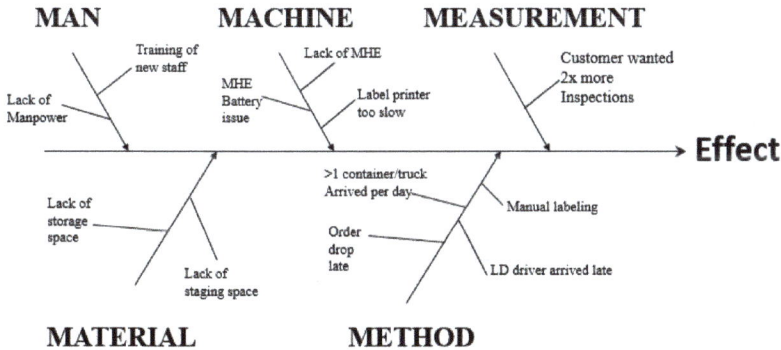

Figure 9. Cause–effect diagram analysis for 10% OT deduction–customer A.

Source: Adapted from a Case Study conducted for one of the SME (Small Medium Enterprise).

- **Root Cause #1**
  - Lack of storage space

- **Symptoms**
  - 43% more volume
  - Overflow to aisle areas
  - Not able to pick efficiently

- **Solution**
  - Excess storage to 2nd floor warehouse
  - Move slow moving SKUs to 2nd floor

- **Timeline**
  - 2 weeks to set up racking
  - 1 week to move slow moving SKUs

Figure 10. Cause–effect solution for one of the root causes.

Source: Adapted from a Case Study conducted for one of the SME (Small Medium Enterprise).

Next, we need to finalize the Kaizen form, which serves as a summary page for our final presentation and a one-page reference document. Following is the partially completed Kaizen form (Figure 13) to demonstrate our approach. It is important to revisit the 5W2H brainstorming ideas and utilize the cause–effect diagram analysis to develop a final version within the established timeline. Consequently, the Kaizen form must be regularly updated to ensure that we produce a document that meets the approval of both the project team and the project lead.

Finally, how can we measure our achievement of a 10% reduction in overtime (OT) or 60 hours through the implementation of countermeasures? There are two types of measurements: lead and lag factors. Lag measures provide results at the end of the month, reflecting what we have accomplished. While it is essential to assess the effects of our countermeasures, relying solely on lag measures means we may be too late to

Minutes of Weekly Kaizen Meeting on: Reduce OT hours at 10% – Customer A

Present:
1. Operation Manager 2. Executive 3. Lead Warehouse Assistant 4. Warehouse Admin 5. Warehouse Assistant

Absent with apologies: N.A

Date	Details	Action By	Deadline
	Weekly action plan updates:		
02-Feb-23	Ops Excellence Team guided team to kick start on Reduce OT hours at 10% as Kaizen project. Therefore, warehouse need to form up a team to kick start this project and the members are: 1. Operation Manager 2. Executive 3. Lead Warehouse Assistant 4. Warehouse Admin 5. Warehouse Assistant	All	16-Feb-23

16-Feb-23	First discussion will be the brainstorming on 5W2H to all team members including the ground staff.
	• Data collection on incoming volume for container and truck
	• Data collection on local distribution driver arrival for pickup
	• Data collection for customer order delays
	• Data collection on current UPH (Unit Per Hour)
	   ○ Receiving (Dock to Stage)
	   ○ Put-away (Stage to Stock)
	   ○ Pick/pack (Pick-Pack to Stage)
	   ○ Releasing of pallet to local distribution
	
	• Check on current pallet location and overflow locations for average and peak periods
	• The next session will provide solutions to mitigate the constraint area.
		Manager Lead Admin Assistant	26-Feb-23

Figure 11.   Weekly Kaizen MoM for weekending (DD-MMM-YY).

Source: Adapted from a Case Study conducted for one of the SME (Small Medium Enterprise).

respond effectively. In contrast, lead measures enable us to predict outcomes before actual results are available. For instance, to reduce accident rates in a warehouse, we might implement lead measures such as hazard reporting to identify and eliminate potential hazards that could lead to minor or serious accidents. In our overtime scenario, we would apply the countermeasures outlined in the Kaizen form by monitoring OT hours daily without compromising customer service. This proactive approach allows us to achieve our desired outcomes by the end of the month, quarter, and year.

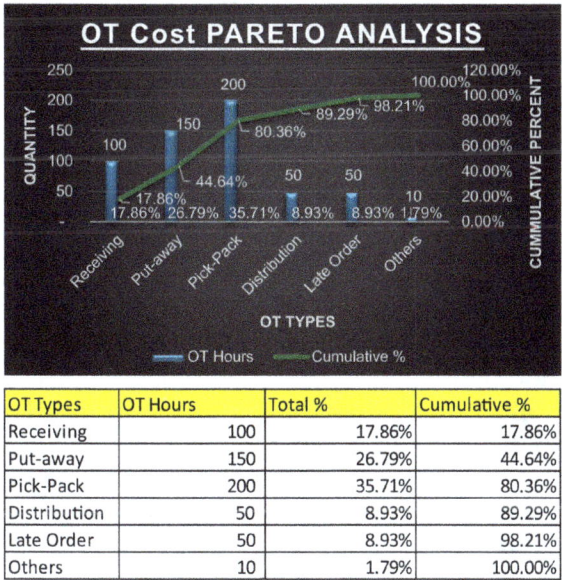

OT Types	OT Hours	Total %	Cumulative %
Receiving	100	17.86%	17.86%
Put-away	150	26.79%	44.64%
Pick-Pack	200	35.71%	80.36%
Distribution	50	8.93%	89.29%
Late Order	50	8.93%	98.21%
Others	10	1.79%	100.00%

Figure 12.   OT cost pareto analysis.

Source: Adapted from a Case Study conducted for one of the SME (Small Medium Enterprise).

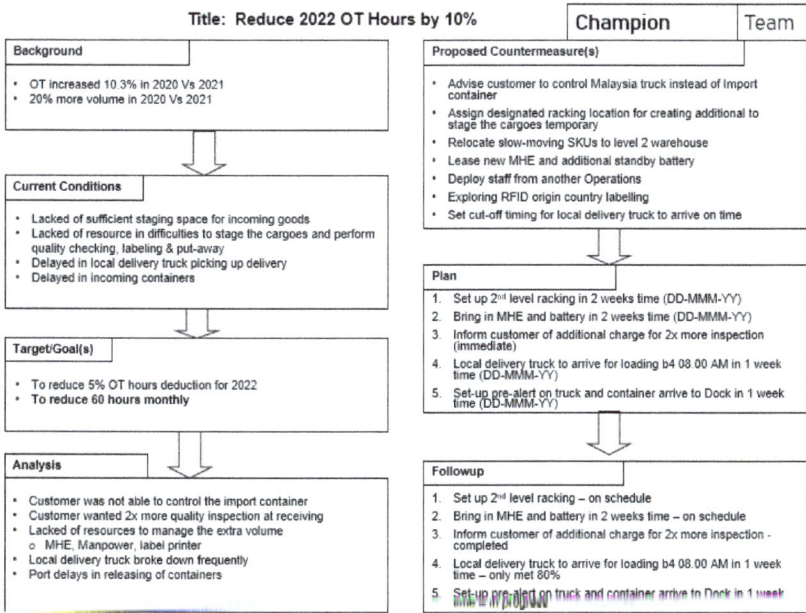

Figure 13.   Partially completed kaizen form for 10% OT deduction.

Source: Adapted from a Case Study conducted for one of the SME (Small Medium Enterprise).

By adopting the approaches outlined above, we can implement continuous improvement in warehouse operations with relative ease. This is largely because you have a deeper understanding of your operations than anyone from outside. However, bringing in someone from another warehouse or department to provide a fresh perspective can also be beneficial.

4. Building a Culture of Continuous Improvement

Establishing a culture of continuous improvement (CI) is essential for organizations aiming to enhance efficiency, innovation, and long-term success. This transformation requires unwavering support from top management, as well as the engagement and empowerment of employees at all levels. By embedding CI principles into daily operations, organizations can unlock their full potential.

4.1 *Leadership commitment and leading by example*

Leaders play a pivotal role in fostering a culture of CI, setting the tone for the entire organization. They must actively communicate the organization's vision, emphasizing the importance of CI and securing employee buy-in. This vision should not only highlight the benefits of CI but also outline specific goals. For instance, Toyota's Production System exemplifies the Kaizen philosophy — a commitment to incremental improvements that, over time, lead to substantial enhancements in performance and efficiency. Similarly, 3M's "15% Culture" allows employees to dedicate a portion of their time to innovative projects, fostering creativity and progress.

Effective leaders go beyond advocacy by leading through action. By participating in CI initiatives and working alongside employees on projects, they demonstrate genuine commitment and model collaboration. This hands-on involvement fosters mutual respect and builds a foundation of trust. Moreover, it equips employees with essential skills in project management and process improvement techniques, reinforcing the value of CI and driving engagement.

Leaders must also create an environment of open communication, where regular meetings and brainstorming sessions encourage employees to share ideas and insights. By actively listening and responding to employee suggestions, leaders can cultivate a mindset of continuous learning and improvement across the organization.

4.2 *Empowering employees*

Empowering frontline employees is critical, as they are directly involved in executing the tasks that drive organizational success. Providing these employees with the right resources — such as practical process improvement tools and targeted training — enables them to contribute effectively to CI initiatives. Clear guidelines and boundaries ensure employees remain aligned with organizational goals, but empowerment should not be a "delegate and forget" approach. Leaders must stay actively engaged through regular follow-ups to provide support and maintain alignment. Recognition plays a key role in sustaining motivation. Celebrating employee achievements, whether through formal awards or informal acknowledgments, validates their efforts and reinforces the importance of CI. For example, implementing quarterly competitions between teams and rewarding the winners with monetary prizes or special events fosters healthy competition and strengthens team cohesion.

4.3 *Training and development*

Continuous learning is a cornerstone of CI. Employees must be trained in process improvement tools that are simple and practical. Introducing these tools through hands-on projects, such as tackling "low-hanging fruit" inefficiencies, demonstrates their applicability and builds confidence. Workshops focusing on best practices in inventory management, warehouse operations, and other critical areas offer valuable insights for improving efficiency. Additionally, leveraging online learning platforms and adopting new technologies help employees stay informed about industry trends, ensuring adaptability and long-term growth.

4.4 *Regular audits and reviews*

To maintain the integrity of established procedures and ensure sustained improvement, regular audits and reviews are essential. These evaluations act as checks and balances, preventing shortcuts and identifying inefficiencies. A dedicated audit team can standardize this process, ensuring comprehensive assessments. For instance, audits conducted by external teams often provide fresh perspectives, uncovering overlooked inefficiencies. Tools such as warehouse audit checklists are invaluable for ensuring consistency and fostering continuous improvement (Figure 14).

Monthly Audit Checklist - Inbound Process

Operation: **Date of Audit:** _____

(1 point each for each matched criteria except otherwise in the allocation)

Related documentation : INBOUND	Score		Findings	Instruction
	Allocation	Obtained		
1 Inbound receiving and putaway processes are established and clearly defined in WI (Work Instruction). - Inspection and accept/reject criteria	5	0		1. Check WI to see if there is a proper procedure for inspection of inbound. (1 point). 2. Observe an incoming container unloading processes to see if workers is following the SOP. (1 point) 3. Check WI to see if there is a reject criteria handbook being observed. (1 points) 4. Check the employee training records for handling of inbound. (1 point) 5. Select one personnel and ask him/her to describe the procedure. (1 point)
2 Identification, segregation, reporting and decision in handling of discrepancy / non-conforming products (both physical and system)	7	0		1. Verify 3 orders of non-conformance products are properly segregated at a designated locations and with right classification . (3 points) 2. Verify selected 3 orders of inbound non-conformance products are indicated in SAP. (3 points) 3. Is the storage location properly label to indicate the non-conforming product? (1 point)
3 Sufficient area allocated or sufficient measure in place for identification for inbound products to prevent mix-up	2	0		1. Check if Inbound Area is sufficient to cater for incomings and with proper holding location before putting away. (1 point) 2. Rule of thumb is incoming and Pick/Pack or waiting for delivery or export products have to be in different pools. (1 point)
4 Check a few locations if the physical putaway location is consistent with the system record, or vice versa	3	0		1.Select 3 locations on the racks and verify in YCH WMS to see if the SKUs and location is correct. (3 points)
5 Verify if the Inbound Damage and Discrepancy are communicated to Versuni on time (within KPI timeframe)	2	0		1. Select 2 recent incidents, one for damages and one for discrepancy. (2 points) 2. Verify that Discrepancy Report for recent incidents are communicated to Versuni on time.
	19 Total:	0		

Figure 14.   Warehouse audit checklist (sample).

Source: Adapted from a Case Study conducted for one of the SME (Small Medium Enterprise).

4.5 Establishing the foundation: Process, policy, and procedure

Before launching CI initiatives, organizations must establish a strong foundation using the 3P Framework: Process, Policy, and Procedure. This framework ensures operational efficiency, safety, and consistency:

- **Process** defines the sequence of steps involved in operations, such as receiving, storing, picking, packing, and shipping goods. Mapping these processes helps identify inefficiencies and streamline workflows.
- **Policy** sets the overarching rules and guidelines for warehouse operations, such as safety protocols and inventory management practices.

Clear policies minimize errors, enhance compliance, and promote a safer work environment.

- **Procedure** provides detailed instructions for executing tasks within a process. Standard Operating Procedures (SOPs) ensure consistency and alignment with organizational goals.

4.6 *Driving efficiency with the flow process chart*

Once CI principles are integrated, organizations can use tools such as the flow process chart to optimize workflows. This chart maps activities such as operations, transport, delays, inspections, and storage, identifying areas of waste. By reducing or eliminating delays and unnecessary transport, organizations can streamline value-added tasks, improve efficiency, and enhance overall productivity.

Establishing a robust culture of continuous improvement in warehouse operations requires a holistic approach encompassing several key elements: leadership support, employee engagement, effective training, technology integration, regular reviews, and recognition of achievements. By embedding these components into the organization's core, warehouses can better adapt to ever-changing demands while continuously improving their overall operational performance.

Incorporating these critical elements not only enhances efficiency but also fosters an environment where every employee feels valued and empowered to actively contribute to the organization's success. This sense of empowerment motivates individuals to take ownership of their roles, share innovative ideas, and collaborate with colleagues to advance continuous improvement initiatives. Ultimately, a culture that prioritizes these aspects creates a dynamic and responsive warehouse operation capable of excelling in today's competitive landscape.

5. Summary

Building a culture of continuous improvement demands an integrated approach that combines leadership commitment, employee engagement, training, empowerment, regular audits, and recognition of achievements. By embedding these principles into the organizational framework, companies can cultivate an environment where employees feel valued and motivated to drive CI initiatives forward. This comprehensive strategy not only improves operational efficiency and fosters innovation but also equips the organization to respond effectively to evolving demands. A well-established culture of continuous improvement ensures sustained success and positions the warehouse for long-term growth and competitiveness in a dynamic market.

Appendix: On-The-Job Training Assessment Checklist

OJT Program: Warehouse Operations—
Operating Forklift (1–5 tonnes)   **Date:** _____

Trainee: _____   **Trainer:** _____

Did the trainee:

		Yes	No
		☐	☐
(1)	Clarify the details of work activities to be carried out with appropriate person based on given work instructions?	☐	☐
(2)	Wear personal grooming and protective clothing in accordance with procedures?	☐	☐
(3)	Select forklift based on capacity and type of load to allow for adequate load weight safety margin?	☐	☐
(4)	Complete the takeover procedures of selected forklift in accordance with requirements?	☐	☐

(5)   Complete pre-operational checks on forklift in accordance with procedures?   ☐   ☐

(6)   Report to the appropriate person any abnormalities in forklift parts in accordance with procedures?   ☐   ☐

(7)   Inspect the safe routes identified and marked for forklift movement?   ☐   ☐

(8)   Inform the appropriate person where preparation for forklift operations cannot be fully completed?   ☐   ☐

(9)   Ensure forklift is mounted at appropriate side using appropriate hand and foot holds?   ☐   ☐

(10)   Correctly adjust operator's seat to ensure safe and positive control of machine?   ☐   ☐

(11)   Conduct all round observation of environment prior to moving off and while operating forklift?   ☐   ☐

(12)   Drive the forklift in chosen direction with smooth and controlled application of all driving controls and ensure his body, limbs, and head are fully within machine confines?   ☐   ☐

(13)   Drive the forklift safely at all times with speeds adjusted in response to environmental conditions and statutory speed limits?   ☐   ☐

(14)   Take appropriate action to stop forklift in an emergency?   ☐   ☐

(15)   Ensure all steering maneuvers are made smoothly and gently with due regard for forklift stability?   ☐   ☐

(16)   Make visual checks on front/rear end swing of forklift?   ☐   ☐

(17) Assess weight and center of balance of loads to be transported?

(18) Carry out stack and de-stack operations in accordance with industry guidelines?

(19) Check the stacks of load for stability and safety and inform appropriate person where stacks are found unsafe or in danger of collapse

(20) Ensure loads to be transported are adequately protected to prevent damage during movement ?

(21) Ensure route of travel is safe and free of obstruction when transporting load?

(22) Adjust fork arms and tilt sufficiently to clear ground?

(23) Drive laden forklift on slope in appropriate direction to maintain stability, traction, adhesion and load security?

(24) Drive forklift on ramps at safe speed, well clear of edge, and without driving across or turning on slope?

(25) Ensure all health and safety requirements are adhered to in operating forklift?

(26) Assess for weight and center of balance of loads to be transported?

(27) Apply positioning and other safety measures to vehicle to be loaded/unloaded in accordance with safety procedures?

(28) Position load on vehicle based on delivery order — position of first load will be last load off?

(29) Position loads in a manner that maintains stability of vehicle and packed tightly together to prevent movement when in transit?

(30) Distribute the total weight of loads evenly from side to side over all axes and not exceed specified safe working load of vehicle?

(31) Carry out loading/unloading activities with due regard to safe stacking height and height restrictions where applicable?

(32) Check dock levelers or bridge plates for safe working load and plac them in appropriate position on vehicle floor?

(33) Query the appropriate person when in doubt about capacities of dock levelers, bridge plates, and lorry floors?

(34) Ensure all health and safety requirements are adhered to when operating forklift?

(35) Park the forklift properly at designated area in accordance with safety procedures upon completion of work?

- Apply hand brake
- Park on level ground
- Free gear
- Position wheel
- Tilt work and lower to ground level
- Switch off engine
- Remove key
- Shut off LPG gas cylinder valve (where applicable)
- Connect charger unit to batteries (where applicable)

(36) Report promptly to the appropriate person where faults are found in forklift operations according to established procedures?

(37) Hand over the forklift in accordance with organizational procedures?   ☐   ☐

(38) Inform the appropriate person where handing over of forklift cannot be fully completed?   ☐   ☐

Result of Performance (circle):   **PASS/FAIL**

Note: Trainee is deemed to have failed the program if one of the above columns is marked 'No'.

Signature of Trainer: _____   **Date:**   _____

Glossary

Logistics and Supply Chain Management

- **3PL (Third-Party Logistics):** Outsourcing logistics services to a third-party provider for supply chain management, including transportation, warehousing, and distribution.
- **Cross-Docking:** A logistics practice where goods are directly transferred from inbound to outbound transportation with minimal storage.
- **Dock Scheduling:** Coordination of loading/unloading schedules to optimize warehouse operations and reduce bottlenecks.
- **Freight Forwarding:** Organizing and facilitating the transportation of goods between destinations, often internationally.
- **Last-Mile Delivery:** The final step of the delivery process where goods reach the end customer.
- **Logistics:** The overall management of the flow of goods, information, and resources from origin to consumption.
- **Multi-Channel Fulfillment:** Managing inventory and delivery across multiple sales channels (e.g. online and brick-and-mortar).
- **Order Fulfillment:** The process of receiving, processing, and delivering orders to customers.
- **Pick and Pack:** Selecting ordered items from inventory and packaging them for shipment.
- **Returns Management:** Handling product returns efficiently, including inspection, restocking, or disposal.
- **Transportation Management System (TMS):** Software to optimize the planning, execution, and tracking of freight movements.

- **X-Docking:** Similar to cross-docking but includes temporary holding in a docking station for better scheduling.
- **Yard Management:** Overseeing the movement and storage of vehicles, trailers, and goods in a warehouse yard.
- **Zone Picking:** Dividing a warehouse into zones where specific pickers are assigned to prepare orders.

Inventory Management

- **ABC Analysis:** Classifying inventory into categories (A, B, and C) based on importance and usage value.
- **EOQ (Economic Order Quantity):** The optimal order quantity that minimizes total inventory costs, including holding and ordering costs.
- **Inventory Management:** The process of ordering, storing, and using a company's inventory efficiently.
- **Just-in-Time (JIT):** A strategy to reduce inventory waste by receiving goods only as they are needed in production.
- **Non-Stock Item:** Items not kept in regular inventory but ordered when required.
- **SKU Rationalization:** Reducing the number of stock-keeping units (SKUs) to simplify inventory management.
- **SKU Standardization:** Ensuring uniformity and consistency in how SKUs are assigned and used.
- **Stockout:** A situation where demand cannot be met due to insufficient inventory.
- **Understock:** Having insufficient stock levels to meet demand, often leading to stockouts.
- **Vendor Managed Inventory (VMI):** Allowing vendors to manage inventory levels based on agreed-upon metrics.
- **XYZ Analysis:** Categorizing inventory based on demand variability, with X being stable and Z highly variable.

Quality and Performance

- **Key Performance Indicator (KPI):** Metrics used to evaluate the efficiency and success of specific business objectives.
- **Quality Control:** Processes to ensure that products meet specified quality standards.

Environmental and Sustainability Initiatives

- **Environmental Management System:** Frameworks for organizations to monitor and improve environmental performance.
- **ISO 14000:** A set of international standards for environmental management practices.
- **Renewable Energy Sources:** Energy derived from sustainable resources, such as solar, wind, and hydro.
- **Sustainable Building Materials:** Materials that minimize environmental impact throughout their lifecycle.
- **Energy Efficiency:** The ratio of useful energy output to total energy input, minimizing waste.
- **Green Warehouse:** An environmentally sustainable facility utilizing energy-efficient practices, renewable resources, and waste reduction strategies.

Costs and Forecasting

- **Anniversary Billing:** A billing cycle tied to a specific annual date, often used for subscriptions or contracts.
- **Demand Forecasting:** Predicting future customer demand using historical data and market trends.
- **Holding Costs:** Costs incurred for storing unsold inventory, including warehousing and depreciation.
- **Ordering Costs:** Expenses related to purchasing and replenishing inventory, such as administrative and delivery fees.

Value-Added Services

- **E-commerce Fulfillment:** End-to-end services for online orders, including picking, packing, and shipping.
- **Value-Added Services (VAS):** Extra services beyond basic logistics, such as kitting, assembly, and custom packaging.

Cross-Docking and Lean Warehousing Practices

- **Cross-Docking:** A logistics strategy that minimizes storage time within a warehouse by facilitating the direct transfer of goods from incoming deliveries to outbound shipments.

- **Lean Warehousing:** An operational strategy that applies lean principles to warehouse processes, aiming to eliminate waste, reduce costs, and improve efficiency.
- **Just-in-Time (JIT):** A methodology that ensures inventory arrives precisely when needed, minimizing storage time and aligning with lean principles.
- **DMAIC:** A structured methodology used in Lean Six Sigma for process improvement, consisting of five phases: define, measure, analyze, improve, and control.
- **Value Stream Mapping (VSM):** A visual tool used to map out the flow of materials and information in a process to identify inefficiencies and opportunities for improvement.
- **Kanban System:** An inventory management system that uses visual cues, such as empty bins, to signal when replenishment is needed.
- **Warehouse Management System (WMS):** A software solution that optimizes warehouse operations, including inventory control, order picking, and space utilization.
- **5S Program:** A lean tool for workplace organization and efficiency, consisting of sort, straighten, shine, standardize, and sustain.
- **Key Performance Indicators (KPIs):** Metrics used to measure the success of specific operational processes, such as order fulfillment speed or inventory accuracy.
- **SIPOC Diagram:** A high-level visual representation of a process, showing suppliers, inputs, processes, outputs, and customers.
- **Inbound Process:** The flow of activities related to receiving, inspecting, and storing goods arriving at a warehouse.
- **Kaizen Bursts:** Specific areas in a process map identified for focused improvement efforts.
- **Fishbone Diagram:** A root cause analysis tool used to identify and categorize potential causes of a problem.
- **Spaghetti Diagram:** A visual tool that tracks the movement of materials or people within a process, identifying inefficiencies and redundancies.
- **Transportation Waste:** Inefficiencies caused by unnecessary movement of goods within a warehouse.
- **Inventory Waste:** Waste resulting from overstocking or inefficient use of storage space.
- **Movement Waste:** Time wasted by employees moving unnecessarily to locate or access items.

- **Waiting Waste:** Delays caused by waiting for resources, such as fork-lifts or space clearance.
- **Defect and Damage Waste:** Waste caused by improper storage or handling, leading to damaged goods.
- **Standard Operating Procedures (SOPs):** Documented procedures that ensure consistency and efficiency in warehouse operations.
- **Dock Appointment Scheduling:** A system for coordinating truck arrivals at a warehouse to streamline the inbound logistics process.
- **Staging Area:** A designated space within a warehouse for temporarily holding goods before they are moved to storage or shipping.
- **Third-Party Logistics (3PL):** Companies that provide outsourced logistics and warehouse management services.
- **Gemba Walk:** A lean tool involving on-site observation to identify inefficiencies and improvement opportunities in a process.
- **Lead Time (LT):** The total time required to complete a process from start to finish.
- **Cycle Time (CT):** The time taken to complete a specific task or activity within a process.
- **Inbound Staging Area:** A temporary holding area for goods that have just been received but not yet moved to their final storage location.
- **As-Is Process Map:** A visual representation of the current state of a process, including all activities and workflows.

Safety and Security

- **WHS:** Workplace health and safety stands for comprehensive framework aimed at ensuring the safety, health, and welfare of workers in various environments.
- **bizSasfe:** The **bizSAFE program** is a Singapore national initiative aimed at enhancing workplace safety and health standards across various industries. It is designed to help companies reduce workplace accidents and injuries while fostering a culture of safety.
- **Risk Assessment:** It is a systematic process used to identify, analyze, and evaluate potential hazards that could negatively impact an organization's operations, employees, or assets. It serves as a foundational element of risk management strategies aimed at minimizing risks and ensuring safety.
- **Control Measure:** There are specific actions or strategies implemented to eliminate, reduce, or manage identified risks in a workplace

or operational environment. These measures are essential for protecting employees, assets, and the overall safety of operations.

- **Hazard Identification:** It is a systematic process aimed at recognizing potential hazards that could cause harm to workers, property, or operations. This process is crucial for maintaining safety and preventing accidents in a setting known for various risks.
- **Risk Evaluation:** It is a critical component of the risk assessment process, where identified risks are analyzed to determine their significance and the appropriate response. It involves comparing estimated risks against established criteria or standards to decide whether they are acceptable or require further action.
- **Risk Management Strategy:** It is a comprehensive plan designed to identify, assess, and mitigate risks associated with warehouse operations. This strategy aims to ensure the safety of personnel, protect assets, and maintain efficient operations while minimizing potential losses.

Index